CREATIVE TOURISM
IN SMALLER COMMUNITIES

SMALL CITIES SUSTAINABILITY STUDIES IN COMMUNITY AND CULTURAL ENGAGEMENT

SERIES EDITORS:

Will Garrett-Petts, Professor of English and Associate Vice-President of Research and Graduate Studies, Thompson Rivers University

Nancy Duxbury Carreiro, Senior Researcher, Centre for Social Studies, University of Coimbra, Portugal, and Co-coordinator of its Cities, Cultures, and Architecture Research Group

Published with the support of Thompson Rivers University.
ISSN 2561-5351 (Print) ISSN 2561-536X (Online)

This series is interested in discovering and documenting how smaller communities in Canada and elsewhere differ from their larger metropolitan counterparts in terms of their strategies (formal and informal) for developing, maintaining, and enhancing community and cultural vitality, particularly in terms of civic engagement, artistic animation, and creative place-making.

No. 1 · *No Straight Lines: Local Leadership and the Path from Government to Governance in Small Cities*
Edited by Terry Kading

No. 2 · *Creative Tourism in Smaller Communities: Place, Culture, and Local Representation*
Edited and with an introduction by Kathleen Scherf

UNIVERSITY OF CALGARY
Press

Creative Tourism in Smaller Communities

PLACE, CULTURE, AND LOCAL REPRESENTATION

EDITED AND WITH AN INTRODUCTION BY

Kathleen Scherf

THOMPSON
RIVERS
UNIVERSITY

Small Cities Sustainability Studies
in Community and Cultural Engagement

ISSN 2561-5351 (Print) ISSN 2561-536X (Online)

University of Calgary Press
2500 University Drive NW
Calgary, Alberta
Canada T2N 1N4
press.ucalgary.ca

LIBRARY AND ARCHIVES CANADA CATALOGUING IN PUBLICATION

Title: Creative tourism in smaller communities : place, culture, and local representation / edited and with an introduction by Kathleen Scherf.
Names: Scherf, Kathleen Dorothy, 1960- editor, writer of introduction.
Series: Small cities sustainability studies in community and cultural engagement ; 2.
Description: Series statement: Small cities sustainability studies in community and cultural engagement ; 2 | Includes bibliographical references and index.
Identifiers: Canadiana (print) 20210133856 | Canadiana (ebook) 20210133953 | ISBN 9781773851884 (softcover) | ISBN 9781773851891 (open access PDF) | ISBN 9781773851907 (PDF) | ISBN 9781773851914 (EPUB) | ISBN 9781773851921 (Kindle)
Subjects: LCSH: Tourism. | LCSH: Tourism—Management. | LCSH: Sustainable tourism. | LCSH: Community life. | LCSH: Community development.
Classification: LCC G155.A1 C74 2021 | DDC 910—dc23

The University of Calgary Press acknowledges the support of the Government of Alberta through the Alberta Media Fund for our publications. We acknowledge the financial support of the Government of Canada. We acknowledge the financial support of the Canada Council for the Arts for our publishing program.

This book is published with financial support from Thompson Rivers University.

Printed and bound in Canada by Marquis Book Printing
♻ This book is printed on Enviro Opaque paper

Cover image: Colourbox 22024929
Copyediting by Ryan Perks
Cover design, page design, and typesetting by Melina Cusano

For Hilda Scherf

Contents

Acknowledgements   IX

Introduction: Creative Tourism in Smaller Communities: 1
    Collaboration and Cultural Representation
        Kathleen Scherf

1   Catalyzing Creative Tourism in Small Cities and Rural   27
    Areas in Portugal: The CREATOUR Approach
        Nancy Duxbury

2   The Interplay between Culture, Creativity, and Tourism   61
    in the Sustainable Development of Smaller Urban Centres
        Elisabete Caldeira Neto Tomaz

3   The Role of Cultural Festivals in Regional Economic   79
    Development: A Case Study of Mahika Mahikeng
        James Drummond, Jen Snowball, Geoff Antrobus,
        Fiona Drummond

4   Creative Yukon: Finding Data to Tell the Cultural   109
    Economy Story
        Suzanne de la Barre

5   When Our Ship Comes In: The Cultural Impact of Cruise   137
    Tourism on Northern Canadian Communities
        M. Sharon Jeannotte

6   Creative Tourism: The Path to a Resilient Rural Icelandic   165
    Community
        Jessica Faustini Aquino, Georgette Leah Burns

7   Placemaking through Food: Co-creating the Tourist   191
     Experience
         Susan L. Slocum

8   *Literary Atlas*: A Digital Resource for Creative Tourism in   209
     Wales
         Kieron Smith, Jon Anderson, Jeffrey Morgan

9   Creative Tourism Opportunities through Film and   229
     Tourism Industry Collaboration
         Christine Van Winkle, Eugene Thomlinson

10   Art Worlds in the Periphery: Creativity and Networking in   259
     Rural Scandinavia
         Solène Prince, Evangelia Petridou, Dimitri Ioannides

Conclusion: Creative Placemaking Strategies in Smaller   283
     Communities
         Greg Richards

Contributors   299

Index   305

ACKNOWLEDGEMENTS

Thanks to Thompson Rivers University, especially the Research and Graduate Studies Office, for granting a 2017–18 sabbatical, during which I carried out most of the early work on this volume, and for funding two editorial assistants, Isabella Cervantes and Sarah Miller, through its Undergraduate Research Apprenticeship Program. Thompson Rivers University is situated on the land of the Secwepemc, within Secwepemc'ulucw, the traditional territory of the Secwepemc People. I am grateful to Universitat Autonòma de Barcelona and Johannes Gutenberg-Universität Mainz for hosting me for one semester each during that sabbatical. My colleague and friend Dr. Nancy Duxbury has been keenly supportive during the entire process of preparing this volume, and I'm delighted to acknowledge her contributions to my work, and to the field of creative tourism in general. I'm pleased that Dr. Greg Richards agreed to contribute the conclusion to the volume when, as any list of references in any study of creative tourism reveals, he is very busy. I am so happy to thank the nineteen other authors who have contributed chapters to this volume, and who have endured numerous emails, outlines, revision requests, questions, and demands for page number citations over the last eighteen months—it has been a pleasure to work with you and, in most cases, get to know a bit about your lives. Isabella Cervantes and Sarah Miller have been very attentive research assistants. My relationship with the University of Calgary has been a long and happy one; I am thrilled to be publishing with U of C Press, and I share its commitment to open access. Ryan Perks and JoAnne Burek provided, respectively, outstanding copy-editing and indexing services—they are unsung heroes both. Thanks to the editors of the Small Cities: Sustainability Studies in Community and Cultural Engagement series, Will Garrett-Petts and Nancy Duxbury, and to U of C Press director Brian Scrivener, as well as to the manuscript reviewers for their helpful suggestions. Finally, to my family, friends, and colleagues, *Ich danke euch allen.*

Creative Tourism in Smaller Communities: Collaboration and Cultural Representation

Kathleen Scherf

Issues arising from overtourism in many of the world's major cities call into question the adage "bigger is better," as do touristic desires for authentic, human-scale immersion in local life, culture, and knowledge. Overtourism accounts for many headlines, and some of these posit an alternate travel experience—for example, Elaine Glusac's 29 August 2019 article in the *New York Times*: "Cooler, Farther and Less Crowded: The Rise of 'Undertourism.'" According to the United Nations World Tourism Organization's (UNWTO) statistics for 2018, the number one consumer trend in tourism is "travel for change: live like a local, seek authenticity and transformation" (5). Overtouristed places are hard-pressed to accommodate this trend, but "undertouristed" places are not. Endogenous community characteristics, tangible and intangible, provide an opportunity for residents and locals to come together to create shared social capital that reverberates in the community. Smaller communities are especially well-suited to hosting tourists who seek connection with the local. While the examples provided in the following chapters fall within most countries' definitions of small or at least medium-size cities, it should be said that population is one way, though not the only way, that we can measure what a "smaller community" is. Such communities might also be viewed as granular—small geographical units bound together by social and cultural networks. These networks are invaluable resources for smaller communities transitioning from a resource-based economy or for places that are otherwise urgently seeking a lifeline to cultural, social, economic, and ecological resilience. Many struggling communities are geographically peripheral, intensifying their need for rebranding themselves as destinations that appeal to tourists who seek to live like a local, so that the locals can keep living in their own location. The ability to provide a sense of place, created by locals with endogenous resources, opens up the possibility of a tourism that leverages the unique cultural characteristics of *any* place, whatever the size or expressed parameters—as long as there are creative tourists who want to engage with it. This situation presents tremendous and exciting development opportunities for smaller communities.

In these communities, which are often reeling from economic insecurity, the ability to create tourism products and experiences is a key strategy for economic diversification, as so many of the chapters in this volume show. The move to identify, promote, and commodify endogenous cultural assets provides not only a wide field for tourists who seek immersion in local, place-based cultures, but also offers the host community the opportunity to carefully and collaboratively map the tangible and intangible characteristics that define its unique fingerprint on the planet and find creative paths to transition from a resource-based economy to one that harnesses its cultural assets to offer place-based experiences to tourists.

The need to save the smaller community lifestyle is urgent: in 2018, the United Nations reported that 55 per cent of the world's population lives in cities, and it projects that this figure will increase to 68 per cent by 2050 (UN 2018). Urbanization is devastating smaller places, demanding the development of alternate economies in order for them to survive. At the same time, with so many people living in cities, smaller communities have the opportunity to exploit their "smallness" and to become attractive tourist destinations. And in megacities, the necessity to create a sustainable tourism industry often places an emphasis on diverting tourists from core "sights" and into local neighbourhoods.

The authors in this book confirm that, if well-grounded in community-oriented people, plans, policies, and practices, innovative development such as that offered by creative tourism initiatives can be ecologically, socially, economically, and culturally sustainable. This volume collects a variety of perspectives on the relationship between creative tourism and smaller communities in an attempt to highlight the importance of a collaborative paradigm in culture-led tourism and to analyze how tourism initiatives can access the creative representation of smaller places. The book offers case studies, descriptions, and critiques of the challenges and possibilities of creative tourism in such communities. It illuminates the distinctive realities of tourism initiatives that are mindfully embedded in particular places, provides an international snapshot of research and practice trends in this area from 2016 to 2019, and reflects a growing interest in tourism in smaller places. In 2020, the disastrous effect of the COVID-19 pandemic on the travel and tourism industry highlights the necessary movement from massive to smaller tourism.

While the book's intent is to provide as wide a perspectival canvas as possible, all chapters were selected with the following question in mind: In what ways are creativity and place-based tourism co-engaged to aid sustainable cultural development in smaller communities? The authors were invited to demonstrate a broad understanding of "smaller communities." They tackle the issue of "creativity" through an array of approaches—and in doing so sometimes invoke the work of Richard Florida (2002), acknowledging his theories' awkward fit with smaller communities that do not have the infrastructure to support a creative class. Particularly in towns reeling from various downturns, including the COVID-19 pandemic, harnessing local creativity is more a matter of economic survival than attracting members of the creative class to live in "cool" neighbourhoods in urban centres.[1]

Six general themes about the relationship of creative tourism and smaller communities emerge in the following chapters: (1) the co-creation by visitor and resident of experiences that feature unique local skills and knowledge; (2) the engagement of visitor imagination by participation in tangible or intangible endogenous culture; (3) the generation or regeneration of sustainable cultural development for the host community; (4) the formation of creative networks to offer touristic experiences; (5) the examination of the processes, policies, and methodologies around creative tourism; and (6) the creative representation of smaller communities. Some chapters, depending on their specific topic, develop more than one of these approaches.

The Centrality of Place in Defining Creative Tourism

As Duxbury and Richards (2019) have recently pointed out, creative tourism—the research field produced by the marriage of cultural and experiential tourism—has evolved dynamically since Richards and Raymond (2000) offered the first definition of creative tourism, which described the experience of tourists exercising their own creative potential by actively engaging with the social and cultural characteristics of the destinations they visit. An alternative to the mass consumption of the tangible cultural heritage of a given destination, creative tourism is tied also to intangible knowledge and skills, which, when shared by locals with visitors, provides an authentic experience unique to the place that hosts it. The tangible and intangible characteristics of place, and the experiential integration of the tourist with them, is the core of creative tourism. For the residents of these places, there is an opportunity to engage with visitors to offer place-based experiences. Because they step "outside the confines of the tourist gaze, cultural and creative tourists are engaging their creative skills to develop new relationships with the everyday life of the destination" (Richards 2011, 1233). And as they do, creative tourism reveals that every place has the ability to attract visitors. Every place can create tourism based on its unique characteristics. This simple fact has changed tourism practice and scholarship. More places are now able to compete for a share of an industry eager to offer authentic and personal travel experiences.

Duxbury and Richards (2019) identify four "overlays" to describe the way the study of creative tourism has evolved in the twenty years since the concept was first introduced (4). As they also point out, the rapid take-up of some of the central ideas of creative tourism, such as an embedded sense of place, the transfer of local knowledge and skills, the interaction of locals and visitors, sustainability, and co-creation, as well as the field's inherent interdisciplinarity, has resulted in a situation where "what is referred to as 'creative tourism' may be linked to very different types of creative activities and creative context, and there is no consistent application of the definitions or terminology relating to creative tourism" (5).

In this volume, we understand "creative tourism" to be an experiential subset of cultural tourism that demonstrates four characteristics: (1) it involves the transfer of culture-based, place-specific endogenous knowledge to the visitor; (2) it includes the experiential participation of the visitor in activities that embody such knowledge; (3) it operates in a collaborative paradigm in some manner; and (4) it demonstrates a longer view beyond the actual tourist experience toward the host community's cultural sustainability. Chapters were selected to represent a range of methodologies and critical approaches that embody this definition, which are enumerated below. First, however, we offer a few words about some of the central concepts in the book: smaller communities, collaborative placemaking, planning processes, identification of cultural resources, and cultural sustainability.

Place and Smaller Communities

We view smaller communities as the smallest geographical unit that can be conceptually bound together by a place-based set of tangible and intangible cultural characteristics that include shared history, heritage, values, traditions, practices, and skills. The residents of a smaller community define its physical and conceptual boundaries. Of course, population size is also relevant. Bonifacio and Drolet (2017) define a "small city" in Canada as having a population under 200,000. In Europe, countries define these communities independently, but mostly small and medium-size urban areas show a population of 5,000 to 250,000 (Eurostat 2018). The communities examined in this volume range from remote, rural settlements to medium-size urban areas. Smaller communities are home to a significant percentage of the national population; for example, according to Statistics Canada, 40.4 per cent of Canadians live in places with a population of less than 99,999 (2016). Eurostat reports the percentage in Europe in 2015 was higher, at 59 per cent (2018). As cited earlier, the UNWTO predicts these per centages will decrease through urbanization.

Greg Richards and Lian Duif (2018) argue that smaller communities may offer many advantages for expressing local culture and embedding creative tourism. Kent Robertson (2001), meanwhile, has enumerated the general features of small cities in the United States, and these features suggest both an attractive alternative to mass tourism as well as a fertile field for connecting with the local culture, because they

- Are human scale, less busy, more walkable;
- Do not exhibit the problems of big cities—congestion, crime, etc.;
- Aren't dominated by corporate presence;
- Lack large-scale flagship or signature projects;
- Have retailing distinguished by independents;
- Aren't subdivided into monofunctional districts;
- Are closely linked to nearby residential neighbourhoods;
- Possess higher numbers of intact historic buildings.

(quoted by Bell and Jayne 2006, 8)

Tourism in global cities also demonstrates the trend toward tourists' engagement with the local, which often translates into connecting with a smaller component of a city.

In *City of Quarters: Urban Villages in the Contemporary City* (2004), David Bell and Mark Jayne argue persuasively for the distinct identity of smaller communities within cities, defined by the daily lives and interests of residents. Some of the communities explored in the following chapters are villages—and sometimes barely that. Some are small and medium-size cities; some are remote, and some are neighbourhoods within a city. The word "smaller" in the volume's title was chosen deliberately because it connotes a sense of measurement against something else; a neighbourhood in a city is a place that is smaller than, and contained by, the larger city around it. Neighbourhoods often replicate the characteristics described by Robertson (2001), and they can respond to Bell and Jayne's (2009) challenge for us to think big about being small. This concept has become very important in urban tourism, especially as more tourists seek that authenticity that is frequently associated with the local (Russo and Richards 2016). A country like Wales is smaller in size and cultural impact than its neighbour, England, for example, and its distinct culture can be a refreshing dive for immersive tourists. Cities are themselves seeking to break down a monolithic concept of how tourists engage with their city, devising strategies to tie visitor experiences to an even more local sense of place, either to provide a more engaged experience with the everyday life of the place in which residents and tourists can interact and share an inter-cultural experience that generates a collaborative sense of placemaking, or to disperse numbers of visitors more widely, out of the central tourist areas, so as to distribute the load across the city and ease the stress on central physical, cultural, and social infrastructure.

Copenhagen's tourism organization, Wonderful Copenhagen, has launched the bold "Localhood" campaign as a strong example of the collaborative placemaking strategy. Proclaiming "The End of Tourism as We Know It," Copenhagen's tourism plan to 2020 speaks to "the experience of temporary localhood":

> Today, fewer and fewer [people] want to be identified as tour-ists. Instead, new generations of travellers seek out experi-ences that not only provide a photo opportunity, but also get their hands "dirty" and immerse them in the destination. The travellers seek out a sense of localhood, looking to experience the true and authentic destination—that which makes a desti-nation unique. With the increasing number of providers and businesses that tap into the sharing and collaborative busi-ness potential, travellers gain increasing access to the local travel experience. (Aarø-Hansen 2017, 5)

In this way, Copenhagen's destination marketing organization hopes to avoid the unsustainable tourism of cities such as Amsterdam and Barcelona, where touristic activity is frequently viewed by locals with approbation, if not outright hostility. Instead, the Localhood campaign asserts that "the delivery of an authentic destination experience depends upon the support of locals, whereas the livability and appeal of our des-tination—and thereby the advocacy of locals—depends on our ability to ensure a harmonious interaction between visitors and locals" (5).

Barcelona's central infrastructure, particularly in the central barrios of El Gótico and El Raval, cannot support the number of tourists flocking there, and the "livability" factor for locals is severely reduced (Goodwin 2016). Barcelona has officially adopted the practice of un-identifying visitors as "tourists" or "outsiders," declaring in its *Barcelona Tourism for 2020: A Collective Strategy for Sustainable Tourism* that

> far from being an outside phenomenon, tourism produces the city and, at the same time, the city shapes tourism's possibilities. Tourism is an inherent and constituent part of the current urban phenomenon. Tourism activities must not be seen as something foreign to the city, they are not "out there," but part of its day-to-day activities, intrinsic dynamics, and even daily life. So tourists do not have to be considered passive players "in the city" but rather as visitors with rights and duties "of the city." (Ajuntament de Barcelona Direcció de Turisme [ABDT] 2017, 7)

Like Copenhagen, Barcelona realizes that "the quality of the tourist experience depends on guaranteeing the well-being of the people who live in the city" (16). In order to relieve the pressure on the central tourist district, Barcelona plans to encourage districts outside the centre to develop sustainable tourism plans (30). As Richards and Marques (2018) point out, Barcelona is not alone in attempting to facilitate such distribution: Amsterdam, Rome, Montreal, and Lisbon have various strategies to do the same. Smaller places, smaller communities, are literally well placed to provide more authentic personal experiences for visitors: they are human-scale.

Sustainable Cultural Development as a Theme in Creative Tourism

Creative tourism and its relationship to sustainable development in smaller communities has become a burgeoning area of study, as the following highlights demonstrate. Early in 2019, José Álvarez-García et al. published a bibliographic review on the subject; spring 2019 saw the publication of Duxbury and Richards's *A Research Agenda for Creative Tourism*, which addresses creative tourism and sustainable development in small and rural communities. In the summer of 2019, Patrick Brouder and Suzanne de la Barre, both at Vancouver Island University, were awarded a prestigious Social Sciences and Humanities Research Council of Canada grant for their project Creative Economies: Exploring the Nexus of Culture and Tourism in Rural and Peripheral Canada (Yukon and Northern BC). In October 2019, CREATOUR, which focuses on sustainable creative tourism in small communities and rural areas, held its third international conference on creative tourism in Faro, in the Algarve region of Portugal. And in November 2019, Duke University Press published *Detours: A Decolonial Guide to Hawai'i*, which reimagines tourist engagement with the island state so that endogenous culture is seen and sustained. The current volume, published in spring 2021, provides an international digest of perspectives on and approaches to engaging smaller communities with sustainable creative tourism.

Sustainable development has most famously been defined by the World Commission on Environment and Development in its so-called Brundtland Report as "development that meets the needs of the present without compromising the ability of future generations to meet their own needs" (1987, 43). As Du Pisani (2006) has pointed out, however, the concept and its definitions were contentiously vague both before and after the publication of the Brundtland Report, also known as *Our Common Future*. The United Nations is more specific: "For sustainable development to be achieved, it is crucial to harmonize three core elements: economic growth, social inclusion and environmental protection. These elements are interconnected and all are crucial for the well-being of individuals and societies" (n.d.).

The rise of the creative economy, and its relation to cultural, experiential, and creative tourism, has placed a spotlight on the role of cultural and creative assets, both tangible and intangible, in culture-led local economic, social, and environmental development initiatives.[2] At a macro level, the UNWTO (2017) has recognized the connection between cultural tourism and the United Nations' Sustainable Development Goals; at a micro level, almost every chapter in this volume speaks to the benefits of collaborative placemaking processes that integrate the possibilities of creative tourism in addressing local sustainability issues. As early as 2005, W. F. Garrett-Petts and Lon Dubinsky argued that "economic development and cultural development cannot be effectively separated, especially in a small city. Tourism promotion, for example, can have only limited success if it is not planned in co-ordination with the cultural sector" (10)

And, as many authors in this volume demonstrate, effective development planning is not only a thematically integrated process, but is also a collaborative community practice.

Smaller communities, although often disadvantaged in terms of their ability to offer tangible resources, do have the ability to generate a bottom-up planning process, tapping into their knowledge of everyday life in their place, which in turn can develop attractive creative tourism experiences based on both tangible and intangible resources. The emphasis on arts-based placemaking, of cultural mapping, of searching out the creative skills and knowledge embedded in the host community, and of developing creative experiences in which visitors can learn about and practise these skills, develops shared social capital in the host destination. Like a creative fingerprint, these experiences are unique to the destination, arising as they do out of the genuine and authentic skills and ways of locals, providing a local authenticity that is not as available to a more homogenized urban city brand. The more the interest of tourists and residents are aligned, the more social cohesion and social capital are generated. Creative placemaking, which Greg Richards addresses in his conclusion to this volume, opens a space for creative tourism as a sustainable development tool. A critical strategy to get there, Richards points out, is "harnessing the creative energy of local communities." Host communities can employ various methodologies and demonstrate various expressions of these features of creative tourism, but all of them demonstrate a reliance on creative people and the organizations and networks they build to engage in creative tourism.

How can a community engage in a collaborative process to identify its sources of creativity? Nancy Duxbury, W. F. Garrett-Petts, and David MacLennan's book *Cultural Mapping as Cultural Inquiry* (2015) offers sixteen different approaches to answering that question, using the following idea as a basis:

> Cultural mapping is regarded as a systematic tool to involve communities in the identification and recording of local cultural assets, with the implication that this knowledge will then be used to inform collective strategies, planning processes, or other initiatives. . . . Together, these assets help define

communities (and help communities define themselves) in terms of cultural identity, vitality, sense of place, and quality of life. (2)

Identifying cultural assets is key to deploying creative tourism and encouraging sustainable development. The 2017 conference Culture, Sustainability, and Place: Innovative Approaches for Tourism Development provided a forum for discussion about "the portrayal of people and places." It recognized the importance of identifying critical themes that emerge when local stakeholders use tourism as a method of local sustainable development; art and culture were seen as key to developing meaningful, sustainable tourism that is place-sensitive and contributes responsibly to sustainable local community development. In their placemaking guide for small cities, Greg Richards and Lian Duif (2018) hope to "inspire a new development agenda for the small city" (3). The basics of the placemaking process, they argue, are meaning, creativity, and resources. In particular, they call attention to intangible resources when they write that "In smaller cities immaterial resources are particularly important, because the culture, creativity, and skills available within the city allows them to make better use of the relatively limited means at their disposal" (18).

For smaller communities especially, community-led economic development that involves cultural mapping can help forge a strong relationship between sustainable development and creative tourism. Identifying creative skills and features of lived experience, defining unique aspects of traditional local heritage, tracing creative networks, developing policies that integrate the needs of stakeholders, and connecting residents and visitors are all strategies that can tightly bind creative tourism and sustainable development. The chapters in this book explore those ideas, and have generated the definition of creative tourism we proposed earlier in this introduction.

Methodologies and Approaches

There are many ways to address our central question (In what ways are creativity and place-based tourism co-engaged to aid sustainable cultural development in smaller communities?). No single edited volume can cover enough ground or offer a sufficient number of case studies to exhaust all possible approaches to this question. We hope, however, that the following chapters will contribute to an understanding of the central question we pose and will generate further study. This volume covers much ground but offers two main approaches to the study of creative tourism: an emphasis on the importance of locally led collaborative placemaking and planning processes in developing sustainable creative tourism enterprises, and the creative representation of place and its role for both the host community and the creative tourist experience in creating cultural capital. These two types of approaches are frequently entwined, and taken together they encompass all the themes embedded in this volume: (1) co-creation by visitor and resident of experiences that feature unique local skills and knowledge; (2) engagement of visitor imagination through participation in tangible or intangible endogenous culture; (3) generation or regeneration of sustainable cultural development for the host community; (4) formation of creative networks to offer touristic experiences; (5) examination of the processes, policies, and methodologies informing creative tourism; and (6) creative representation of smaller communities.

Tourism, Development, Policy: The Collaborative Turn of Creative Tourism

In chapter 1, Nancy Duxbury explains and reflects on the integrated approach of CREATOUR (Creative Tourism Destination Development in Small Cities and Rural Areas), a three-year project that, aligned with Tourismo Portugal's desire to redistribute tourism from Porto, Algarve, and Lisbon, is headquartered at the University of Coimbra's Centre for Social Studies, where Duxbury serves as CREATOUR's principal investigator. CREATOUR's massive undertaking examines, in research and practice, how to build a sustainable creative tourism sector. Forty organizations across the country were selected as pilots to develop creative tourism experiences/products using endogenous resources, both tangible and intangible, in projects designed and implemented by locals. These pilots are connected to university research centres, and together they comprise a research-and-application network linked through an advisory council, conferences, meetings, workshops, and publications. This integrated approach provides a framework for developing tourism products that connect visitors and locals through creative self-expression, develops and shares best practices, and provides a living laboratory for researchers to study the creation of sustainable cultures and economics in smaller communities through tourism. CREATOUR has recently expanded, adding a branch in the Azores. In one way, we can view this project as a laboratory that conducts experiments and discusses processes and results in close collaboration between research and practice.

In chapter 2, Elisabete Tomaz examines the role of creative activities as drivers of local development, noting that tourism—especially cultural tourism and its close ties to creative tourism—can play an important role in culture-led development. The role of culture in the development of large urban centres is well-documented, and she believes the same dynamic can apply in small and medium-sized cities. She offers five European examples from the Czech Republic, England, Finland, Portugal, and Italy. Tomaz explains how tourism activity has a relevant place in urban development strategies, not only economically but also by highlighting and supporting local identity and culture. The integration of tourists and residents, a defining characteristic of creative tourism, can strengthen local creative and innovative capacity, resulting in a more sustainable development program.

In some ways, the planning process examined in chapter 3 shares the goals of the cases Tomaz examines, except that it highlights a top-down example of tourism planning that is short on local collaboration. Here, James Drummond, Jen Snowball, Geoff Antrobus, and Fiona Drummond use the South African Cultural Observatory Framework for the Monitoring and Evaluation of Publically Funding Arts, Culture, and Heritage (2016) to assess the impact of the Mahika Mahikeng Music and Cultural Festival, which takes place in the capital of South Africa's North West province every December. The role of festivals in local economic development and place- and identity-making is well-documented, as the literature review in chapter 3 shows. As the authors point out, this festival is a government initiative that

> seeks to capitalize on the development potential of the cultural and creative industries and cultural tourism and is linked to the "Mahikeng Rebranding, Repositioning & Renewal Programme" of the provincial government. The goals of the Festival include promoting cultural and heritage tourism;

celebrating artists in the region and nation; repositioning and rebranding Mahikeng and the North West province as a cultural hub; stimulating economic growth; and creating jobs in the music and cultural industries.

Interestingly, the authors' evaluation of the festival indicates that it may be a little *too* rooted in place, as the audience surveys suggest that Batswana regional culture was overemphasized, with 78 per cent of attendees identifying Setswana as their home language. So, while endogenous cultural resources were definitely expressed and shared, the economic impact of attracting "new dollars" from visitors (the neighbouring countries of Namibia and Botswana were targeted) was less successful. While the Mahika Mahikeng Festival did attempt to employ culture-led community economic development in the region, there is little evidence that cultural interactivity was successful in this enterprise. The chapter offers an example of a creative tourism opportunity that was missed, arguably attesting to the importance of the collaborative paradigm and embedding a meaningful relationship with locals in developing creative tourism.

In chapter 4, Suzanne de la Barre illustrates how cultural mapping is essential to the process of placemaking. Using the Yukon as an example, she argues for the necessity of uncovering and using data at the local level to understand the cultural sector, and its relation to creative tourism, in northern peripheral regions. Spatial and cultural representation, including Indigenization, are vital placemaking elements; shared with stakeholders, these elements in turn should inform any policy aimed at development, including tourism. Any creative tourism experiences must be embedded in the information provided by such data. As she explains in her chapter, de la Barre is concerned with "the impact the cultural sector is having on the territory, including contributions to resident quality of life and community well-being, as well as the engagement between the cultural and tourism sectors." Especially in such large but sparsely populated northern areas, the author points to the crucial role of networks in collaborative placemaking processes that consider nature, tourism, funding support, cultural capital, multiculturalism, and Indigeneity. Weaving tangible and intangible community assets through a community-based development plan, including creative networks among the stakeholders, is key to sustainable development. The provision of creative tourism experiences can be a strategic contribution to the building of resilient communities in economically challenged, peripheral regions with small populations. De la Barre argues that in order to build such sustainable development plans, local data must be mined and considered. Remote northern communities are not only fragile in terms of their physical environments, but also in their social and cultural contexts, given their peripherality, small populations, distance between communities, and, most importantly, threats to Indigenous world views and practices.

We remain in the North for chapter 5, in which M. Sharon Jeannotte examines the issues that arise when tiny and fragile Arctic communities are confronted with mass tourism; she refers directly to the August 2016 entrance of the thousand-person cruise ship *Crystal Serenity* into Canada's Northwest Passage. To this end, her chapter "explores risks to environmental, economic, social, and cultural sustainability in these communities, as well as the complicated relationship between creative practices, cultural tourism, Indigenous/Inuit values and world views, and local planning practices in this part of Canada." Crucially, her chapter also examines "Some of the ethical and practical implications of Inuit/ cruise passenger interactions, and explore[s] the adaptive capacities of these communities to cope with larger and more frequent cruise tourism incursions."

Jeannotte's sensitive and broad study of the complicated web of factors that need to be considered in planning and implementing creative tourism in this small community concludes with three strong and well-reasoned recommendations. Especially in the North, visitors and creative tourism providers must work to co-create a sense of place with an "ethic of care for visited places" (Walker and Moscardo 2016, 1246). Jeannotte's chapter calls attention to the consideration of "last chance tourism," particularly its presence in polar destinations as a result of global climate change. Her concerns have been recently and grimly echoed by Groulx et al. in *Annals of Tourism Research* when the authors write that "last chance visitors extract emotional benefit and an enhanced sense of self by developing a connection to a vulnerable, iconic environment, but remain resistant to making decisions that sustain the source of this relationship beyond their individual experience" (Groulx et al. 2019, 210). Jeannotte's chapter speaks to the necessity of a local collaborative paradigm in developing sustainable creative tourism policy that preserves cultural and natural heritage in fragile smaller communities.

In chapter 6, Jessica Aquino and Leah Burns view creative tourism through the lens of creative resilience in the small village of Húnaþing vestra, a municipality of around twelve hundred inhabitants located in the northwest of Iceland, where the population is slowly declining. Given the size and the peripherality of their location, the residents, as the authors document, have employed creativity as a form of resilience, pushing through the industrialization of the fishing, sheep, and dairy industries and looking toward experiential and creative tourism in order to sustainably develop their community. Aquino and Burns focus on the case of the Icelandic Seal Center, established in 2005 "by a group of locals who recognized an interest in seal-watching tourism but who also wanted to develop sustainable community tourism for Húnaþing vestra." This collaborative planning effort, which not only produced a successful tourism product, but has also served as a community-building, placemaking project, and has led to a revitalization of the entire region. As part of this process, the local creative industries were boosted as stakeholders, stressing the importance of their involvement in this community-based tourism initiative. The authors argue that their chapter showcases a fine example of a sustainable, embedded, and locally developed place-based creative tourism destination.[3]

Creative Expression of Place: The Artistic Turn of Creative Tourism

As Duxbury and Richards (2019) note, not all definitions of creative tourism include specifically artistic interventions and engagement as a necessary condition, but many practices and studies of creative tourism do develop such experiences. The remaining chapters in this volume examine the role of artists in creative tourism, demonstrating how crucial artistic networks are, especially in smaller and geographically dispersed locations, to establishing a sustainable cultural environment that can contribute to economic sustainability, influence policy through culture-led economic development, and, finally, cultivate creative tourism opportunities. Artistic products are tangible expressions of cultural heritage; the skills and knowledge involved in artistic production are intangible and endogenous aspects of culture that are transferable to visitors, and, when actively sought out by creative tourists, they can energize, regenerate, and strengthen the representation of place identity. In all cases, a process of cultural mapping is seen as a foundational strategy to creative place-making, which Greg Richards explores in the book's conclusion. These chapters in this volume shine a light on the importance of clusters and networking—another vital aspect of the collaborative paradigm in creative tourism.

Culinary arts and their attendant traditions represent place, and they have of course held a prominent position in tourism. As Lee et al. have suggested, food clusters are important constituents of place-based creative tourism in smaller communities (2016). The provision of interactive culinary experiences has become a staple of creative tourism; cooking classes abound. Susan Slocum argues persuasively in chapter 7 that more scholarship is needed on local food tourism as a cultural development tool for smaller communities. Rather than visitors symbolically and literally consuming the local culture through making or just eating its food, regardless of food's role as a signifier of place, Slocum suggests that attention to "foodways"—the cluster or network of activities that involve cultural decisions at every step of the food system—via co-created experiences that allow tourists to explore the culture that produces the food traditions of the region is a key to deploying food tourism as a tool of sustainable development. Her approach is echoed by other scholars, such as de la Barre and Brouder on placing food in the Arctic tourism experience (2013). For example, Sangkyum Kim et al. (2019), in their discussion of Mizusawa udon, suggest that visitors appreciate "the traditional, historical and cultural values attached to the Mizusawa udon noodle as a form of Japanese intangible food heritage. Despite the ordinary nature of udon noodle in the Japanese context, Mizusawa udon noodle holds a unique status associated with foodways of the region's identity which has transformed into a commodity for tourists" (183). Slocum's chapter, likewise, "calls for a broader approach to food tourism studies as a means to reflect on the shared construction of food experiences, and situates that approach as a potential method for creative tourism development and agent of change."

In chapter 8, Kieron Smith, Jon Anderson, and Jeffrey Morgan's cultural mapping project offers a strikingly innovative approach to creative tourism by deep mapping the literary creativity of twelve physical and social spaces in the small nation of Wales. In their chapter, they describe their project, *Literary Atlas: Plotting English-Language Novels in Wales*,[4] an interdisciplinary literary geography project that plots twelve anglophone Welsh novels on a digital map of Wales. The project seeks to use digital technology to engage visitors with the local as it is understood by novelists. The deep map format provides rich content for each novel as well as its location and tangible and intangible heritage. And as the authors explain in their chapter, "alongside this, it has designed an evaluative framework, organizing reading groups and field trips to locations around Wales, encouraging participants—including tourists—to engage with Welsh space and place through the lens of literature, while at the same time evaluating these engagements as potential uses of the *Literary Atlas* website."

The use of digital media as a tool to express the complex process of place creation, and to engage visitors in an authentic sense of the distinctive and unique features of the "place" in a mediated setting, is timely in an age of digitally enabled tourism—especially in Wales, which the authors argue could be seen as a smaller national community, in that its literary heritage is often overlooked in favour of its larger neighbours in the United Kingdom. One section of the *Literary Atlas* website allows users to upload their own five-hundred-word "microfictions," which can be any form of creative literary reflection or relevant photos, by pinning them on a location they select on a digital map. In this way, using the interactive tools offered by digital media, visitors can be part of the collaborative composition of Welsh places. As Clifford McLucas (2000) points out, deep maps create a space in which the professional and the amateur can co-create a place, as they do in the *Literary Atlas* project. Visitors can express, via their literary contributions, their engagement with the small places of Wales. Studies of smart tourism would do well to examine this novel initiative.

In chapter 9 we move from literary creative tourism experiences to film. Christine Van Winkle and Eugene Thomlinson explore the opportunities for the film and television industries to benefit both their projects and the small communities in which they film through creative tourism planning and development. Their research uncovers three ways in which the film and television industries could collaborate with tourism destination management organizations for mutual benefits in smaller communities, all of which speak to the collaborative paradigm in creative tourism: marketing and communication, advocacy and lobbying, and resource sharing and development. They argue that this development strategy can enhance the rising interest in film tourism: "both film and tourism rely on the communities in which they exist. When community members reap the benefits of film-induced tourism they are more like to offer valued experiences to their guests." Van Winkle and Thomlinson call for collaboration among stakeholders in a community-based planning process, and they underscore the need for cultural mapping to reveal and identify creative clusters that can be leveraged to produce film-related creative tourism. The case studies included in this chapter provide examples of how such collaboration does and does not work. The authors point out that, while most smaller communities presently realize only short-term benefits from most film-related creative tourism, collaboration between industries in a community of creatives can nurture a more sustained creative tourism practice. Van Winkle and Thomlinson end with a strong call for more research into this possibility.

The theme of creative clusters in rural environments is highlighted in chapter 10. Solène Prince, Evangelia Petridou, and Dimitri Ioannides make the point that creativity in relation to economic growth has typically been studied from the point of view of larger cities. By contrast, in their chapter they "offer insights to the development of such arrangements and theories that demonstrate enhanced sensitivity to the nature of creativity and networking in rural areas, especially in light of these regions' peripherality, where reliance on public funding and tourism development are commonly used to counter limited economic and social opportunities." With the decline of primary industries, rural communities have had to reinvent themselves as sites of touristic consumption, relying on the cultural assets they have to attract creative tourists. The authors examine the importance and role of embedded artistic clusters in establishing a strong sense of territorial cohesion and sense of place, especially in local development schemes. They ask: How are creative clusters formed in peripheral places, such as the two small areas in Scandinavia that provide the basis for their research—the Danish island of Bornholm and the sparsely-populated county of Jämtland in Sweden? How can these clusters contribute to making places that are attractive to residents and visitors? They conclude that artist networks have the potential to enhance life in peripheral locations as creative domains, to enjoy a greater voice in collaborative planning, and to play a significant role in co-placemaking.

Greg Richards concludes the volume with a review of the themes covered in each chapter and a consideration of placemaking and its role in creative tourism in smaller communities. He finds that the small size of the communities examined here can be both a challenge and an advantage. A number of the chapters comment on the comparative manageability of smaller communities, perhaps enabling a smoother collaborative placemaking process. Richards observes that "if a place is good to live in, it will also be good to visit," and he argues for a placemaking mindset based in the emotional connection and meaning of the place for both residents and the visitors that seek to engage with and immerse themselves in a destination's local identity or "sense of place." The "everyday creativity" of the local, Richards argues, is the touchstone for creative placemaking, and the source of creative tourism opportunities in smaller places. Many chapters in this volume reference smaller communities' ability to engage in bottom-up creative placemaking processes, which in turn often spawn creative tourism ideas.

Creative Tourism: Conditions for Sustainable Cultural Development

The volume's chapters, taken together, and in the context of the definition of creative tourism suggested in this introduction, identify five interrelated circumstances that, when present and active, provide favourable conditions that enable creative, place-based tourism to contribute to sustainable cultural development in smaller communities.

First, the host community recognizes and promotes its embedded sense of place, its history, and its tangible and intangible cultural resources, offering the authenticity that creative tourists desire through experiences that provide opportunities for visitors to consciously engage themselves with the endogenous knowledge, skills, traditions, and processes that define the culture of that place. Chapters specifically addressing the condition of asset recognition include Duxbury on CREATOUR, de la Barre on identifying cultural data, Aquino and Burns on the collaborative creation of the Icelandic Seal Center, Slocum on recognizing local foodways and developing their creative tourism potential, Smith, Anderson, and Morgan on the cultural mapping process for *Literary Atlas*, and Richards on asset identification and local placemaking, and their role in supporting healthy, vital smaller communities with robust creative tourism opportunities, which in turn feed the cultural well-being of the host community.

Second, creative tourism initiatives are more sustainable and contribute to a resilient culture when cultural assets have been determined collaboratively through a culture- and/or community-led planning process. The recognition of tangible and intangible place-based cultural assets, and the energetic regeneration of the skills and knowledge associated with attracting, involving, and weaving together both residents and visitors, vitalizes and sustains the host destination by highlighting its unique sense of place as reflected in its culture. In her chapter, Tomaz describes how this process has functioned (or not) in small and medium-size urban centres in Europe; both Drummond et al. and Jeannotte take a critical approach to the need for such collaborative planning in their chapters, as does Richards in his conclusion. Van Winkle and Thomlinson show how planning involving stakeholders in the film industry and film locations could open up a more robust and sustainable creative tourism opportunity. De la Barre explains how using cultural data in a culture-led development process can help foster creative tourism projects in communities. Not only does such a process harvest local ideas and enthusiasm, it also embeds residents in the initiatives discussed, and creates a sense of ownership and shared success. The chapters in this volume speak to the benefit of a collaborative, community-based, culture-led planning process, not only as a sustainable development tool for the community, but also as a generator of creative tourism enterprises. Furthermore, the provision of these sorts of enterprises serves to reinforce local and unique ways of life and strengthen the longevity of endogenous cultural resources.

Third, because smaller communities are often disadvantaged by a lack of density and the dispersed co-location of players in the cultural and creative sectors, it is especially important for host communities who hope to benefit from creative tourism to support extant creative clusters or networks or to create the circumstances in which they could grow. In these smaller nodes of creativity, often physically isolated or geographically peripheral, it is necessary for social and cultural networks to band together, not only for sharing resources and for co-promotion, but also to speak with a unified voice to policy-makers, planners, and funders. The establishment of networks of creative practitioners and supporters in smaller communities is a pivotal aspect of sustainable cultural development when deploying creative tourism as a tool of cultural resilience. In their chapter, Prince, Petridou, and Ioannides demonstrate how such clustering has benefitted creative tourism in the peripheral regions of Scandinavia; Van Winkle and Thomlinson advise that it could work well in smaller communities that provide film locations; as Duxbury shows, creating networks is one of the main concerns of CREATOUR, and the project also emphasizes the benefits of connecting the academic, cultural, and tourism stakeholder sectors. Champions for a particular creative tourism idea, such as those involved in the Icelandic Seal Center described by Aquino and Burns, must get other residents on board, join with other sympathetic networks, or create those networks. While top-down ideas can work (but sometimes not as successfully as envisioned, as Drummond et al. demonstrate in their chapter), creative tourism initiatives that recognize or develop cultural networks and clusters are embedded in the history, culture, and knowledge of a particular place; they are developed by committed and enthusiastic locals, and create genuine pride in genuine place.

The identification of cultural assets through collaborative and inclusive planning processes that involve and support artistic, cultural, and creative networks, and that have the support of local stakeholders, provide the necessary setting for the fourth condition in which creative tourism can flourish and contribute to sustainable cultural development: the relationship between visitors and residents. The collaborative paradigm we suggest as one of the defining components of creative tourism includes not only collaboration among the host community stakeholders, but also between tourists and locals. Or rather, we might say, along with Barcelona's destination management organization, that visitors *are also* host community stakeholders. A respectful and energetic connection not only fosters the ability to, as Richards says, "co-create place," but also makes real the ideal of visitors engaging in the everyday life of the place, thus sustaining the place-based culture, developing it, and vitalizing it by contributing to the destination culture their own experiences of it, shared with residents. Together, they create place. Without the participation of locals, creative tourists will be disappointed in their search for connecting with the everyday life of the host community.

But to sustain a creative tourism sector, these four conditions require a fifth, the one necessary condition that is a key to the entire endeavour: the creative tourist. Duxbury and Richards have argued that insufficient profiling of the creative tourist represents a gap in creative tourism scholarship, and they have called for future research in the area (2019, 179). The authors in this volume stress the need for a collaborative paradigm involving stakeholders, including tourists; the chapters seem to call for visitors who wish to participate in experiences associated with a host community's artistic and cultural heritage, who recognize their agency as contributors to cultural sustainability, and who desire to take up a place in a culture-based collaborative paradigm. This may require a new mindset for being a *visitor*. Tucking into a local culture in this way as a visitor can only happen when the participative arena is manageable—or, put another way, when it is small. As the tourism industry struggles to recover from calamitous reality of *Coronazeit*, we might find that small is the new big.

NOTES

1   For a solid historical analysis and critique of the proliferation of Florida's creative cities strategy and how it relates to small cities in British Columbia, see Karsten (2018).

2   See Dessein et al. (2015) and UNESCO (2016).

3   For more discussion on approaches to co-creation in communities, see the special issue of Tourism Recreation Research, 44.3, 2019, "Critical Issues in Tourism Research," edited by Giang Phi and Dianne Dredge.

4   The project's website is available at http://www.literaryatlas.wales/en/.

References

Aarø-Hansen, Mikkel. 2017. *The End of Tourism as We Know It.* Copenhagen: Wonderful Copenhagen (DMO). http://localhood.wonderfulcopenhagen.dk.

Aikau, Hokulani K., and Vernadette Vicuna Gonzalez, eds. (2019). *Detours: A Decolonial Guide to Hawai'i.* Durham, NC: Duke University Press.

Ajuntament de Barcelona Direcció de Turisme. 2017. *Barcelona Tourism for 2020: A Collective Strategy for Sustainable Tourism.* Barcelona: Impremta Municipal.

Álvarez-García, José, Claudia Patricia Maldonado-Erazo, María de la Cruz del Río-Rama, and María Dolores Sánchez-Fernández. 2019. "Creative Tourism in Small Cities and Rural Areas: A Bibliographic Review." *Enlightening Tourism* 9, no. 1: 63–94.

Bell, David, and Mark Jayne, eds. 2004. *City of Quarters: Urban Villages in the Contemporary City.* Aldershot, UK: Ashgate.

———, eds. 2006. *Small Cities: Urban Experience Beyond the Metropolis.* New York: Routledge.

———. 2009. "Small Cities? Toward a Research Agenda." *International Journal of Urban and Regional Research* 33, no. 3: 683–99.

Bonifacio, Glenda, and Julie Drolet. 2017. *Canadian Perspectives on Immigration in Small Cities.* Cham, CH: Springer Nature.

Centro de Estudos Sociais, Universidade de Coimbra. 2017. "Culture, Sustainability, and Place: Innovative Approaches for Tourism Development." Conference website; event held in Ponta Delgada, Azores, Portugal, 11–13 October 2017. Accessed 25 February 2019. https://www.ces.uc.pt/eventos/culture_sustainability/.

de la Barre, Suzanne, and Patrick Brouder. 2013. "Consuming Stories: Placing Food in the Arctic Tourism Experience." *Journal of Heritage Tourism* 8, nos. 2–3: 213–23.

Dessein, Joost, Katriina Soini, Graham Fairclough, and Lummina G. Horlings, eds. 2015. *Culture in, for and as Sustainable Development: Conclusions from the COST Action IS1007 Investigating Cultural Sustainability.* Jyväskylä, FI: University of Jyväskylä.

Du Pisani, Jacobus A. 2006. "Sustainable Development—Historical Roots of the Concept. *Environmental Sciences* 3, no. 2: 83–96.

Duxbury, Nancy, ed. 2021. *Cultural Sustainability, Tourism, and Development: (Re) Articulations in Tourism Contexts.* London: Routledge.

Duxbury, Nancy, and Greg Richards, eds. 2019. *A Research Agenda for Creative Tourism.* Cheltenham, UK: Edward Elgar Publishing.

Duxbury, Nancy, and Kathleen Scherf. 2018. Final report on international conference *Culture, Sustainability, and Place: Innovative Approaches for Tourism Development.* Coimbra: Centro de Estudos Sociais, Universidade de Coimbra.

Duxbury, Nancy, W. F. Garrett-Petts, and David MacLennan. 2015. *Cultural Mapping as Cultural Inquiry.* New York: Routledge.

Eurostat. 2018. *Eurostat Regional Yearbook 2018*. Luxembourg: Publications Office of the European Union.

Florida, Richard. 2002. *The Rise of the Creative Class. And How It's Transforming Work, Leisure and Everyday Life*. New York: Basic Books.

Garrett-Petts, W. F., and Lon Dubinksy. 2005. "Working Well, Together: An Introduction to the Cultural Future of Small Cities." In *The Small Cities Book: On the Cultural Future of Small Cities*, edited by W. F. Garrett-Petts, 1–13. Vancouver: New Star.

George, Richard, and Irma Booyens. 2014. "Township Tourism Demand: Tourists' Perceptions of Safety and Security." *Urban Forum* 25: 449–67.

Glusac, Elaine. 2019. "Cooler, Farther and Less Crowded: The Rise of 'Undertourism.'" *New York Times*, 29 August. https://www.nytimes.com/2019/08/29/travel/colorado-overtourism.html.

Goodwin, Harold. 2016. "OverTourism: What Is It and How Do We Address It?" *Responsible Tourism Partnership*, 27 October. https://responsibletourismpartnership.org/overtourism/.

Groulx, Mark, Karla Boluk, Chris Lemieux, and Jackie Dawson. 2019. "Place Stewardship among Last Chance Tourists." *Annals of Tourism Research* 75: 202–12.

Karsten, Sharon. 2018. "Utopian Visions of Small City Transformation." PhD diss., Simon Fraser University.

Kim, Sangkyum, Eerang Park, and David Lamb. 2019. "Extraordinary or Ordinary? Food Tourism Motivations of Japanese Domestic Noodle Tourists." *Tourism Management Perspectives* 29: 176–86.

Lee, Anne H. J., Geoffrey Wall, Jason F. Kovacs, Sook Young Kang. 2016. "Food Clusters and Creative Tourism Development: A Conceptual Framework." *Journal of Rural and Community Development* 11, no. 2: 72–88.

McLucas, Clifford. 2000. "Deep Maps: For the Three Landscapes Project." Accessed 1 July 2015. http://cliffordmclucas.info/deep-mapping.html.

Phi, Giang, and Dianne Dredge. 2019. "Critical Issues in Tourism Co-creation." *Tourism Recreation Research* 44, no. 3. https://doi.org/10.1080/02508281.2019.1640492.

Richards, Greg. 2011. "Creativity and Tourism: The State of the Art." *Annals of Tourism Research* 38, no. 4: 1225–53.

———. 2014. "Creativity and Tourism in the City." *Current Issues in Tourism* 17, no. 4: 119–44.

Richards, Greg, and Lian Duif. 2018. *Small Cities with Big Dreams: Creative Placemaking and Branding Strategies*. London: Routledge.

Richards, Greg, and Lénia Marques. 2018. "Creating Synergies between Cultural Policy and Tourism for Permanent and Temporary Citizens." Barcelona: United Cities and Local Governments.

Richards, Greg, and Crispin Raymond. 2000. "Creative Tourism." *ATLAS News* 23: 16–20.

Roberts, Les. 2016. "Deep Mapping and Spatial Anthropology." *Humanities* 5, no. 1: 1–7.

Robertson, Kent. 2001. "Downtown Development Principles for Small Cities." In *Downtowns: Revitalizing the Centers of Small Urban Communities*, edited by Michael A. Burayidi, 9–22. Routledge: New York.

Russo, A. Paolo, and Greg Richards. 2016. *Re-inventing the Local in Tourism: Producing, Consuming and Negotiating Place*. Bristol, UK: Channel View Publications.

Till, Karen E., ed. 2010. *Mapping Spectral Traces*. Blacksburg: Virginia Tech College of Architecture and Urban Studies.

UNESCO. 2016. *Culture: Urban Future. Global Report on Culture for Sustainable Urban Development*. Paris: UNESCO.

United Nations. n.d. "The Sustainable Development Agenda." Accessed 11 March 2019. https://www.un.org/sustainabledevelopment/development-agenda/.

United Nations, Department of Economic and Social Affairs. 2018. Accessed 5 October 2019. https://www.un.org/development/desa/en/news/population/2018-revision-of-world-urbanization-prospects.html.

UNWTO (United Nations World Tourism Organization). 2017. *Tourism and Culture Synergies*. Madrid: UNWTO.

———. 2019. *International Tourism Highlights, 2019 Edition*. Madrid: UNWTO. https://doi. org/10.18111/9789284421152.

Walker, Kaye, and Gianna Moscardo. 2016. "Moving Beyond Sense of Place to Care of Place: The Role of Indigenous Values and Interpretation in Promoting Transformative Change in Tourists' Place Images and Personal Values." *Journal of Sustainable Tourism* 24, no. 8/9: 1243–61.

World Commission on Environment and Development. 1987. *Report of the World Commission on Environment and Development: Our Common Future* [Brundtland Report].

Catalyzing Creative Tourism in Small Cities and Rural Areas in Portugal: The CREATOUR Approach

Nancy Duxbury

Introduction

In recent years, tourism in Portugal has grown exponentially; indeed, it is currently one of the main drivers of the Portuguese economy. Yet while all regions of the country report visitors and offer various "attraction" activities, historic sites, and beautiful locales, tourism still remains heavily concentrated in the large cities of Lisbon and Porto as well as the traditional beach-and-sun Algarve region. Growing concern (internationally) over the negative impacts of overtourism and tendencies toward tourism homogeneity at a time when travellers are increasingly seeking meaningful and authentic experiences loom over this picture. At the same time, from a domestic perspective, finding sustainable development options and possibilities for smaller communities in the interior, especially those that are remotely situated, is an ever-present concern. Turismo Portugal, the national tourism agency, has been making statements about the desire to pull tourists away from overvisited areas and redistribute them to other regions. This could provide a possible opportunity for many smaller communities if attractive offers can be designed and communicated to appropriate niche markets, and if these initiatives are developed with an eye to accentuating the quality of life for local residents, maintaining local control, and designing for local benefit.

Creative tourism is commonly described as a reaction to the growing mass marketization of cultural tourism mixed with the growing desire of travellers to play more active roles in their journeys. Between these two dynamics, however, the question of how to catalyze and develop a creative tourism "sector"—especially in non-metropolitan contexts—is rarely addressed in the creative tourism literature. Creative tourism is an internationally emerging niche model of tourism development that has been steadily evolving in geographically and culturally diverse contexts (Duxbury and Richards 2019), but there are still many gaps in our knowledge about creative tourism development, its evolutionary dynamics, and strategies for sustainable approaches. With this as its context, the CREATOUR project brings together teams in five research centres and forty pilot organizations to promote, learn, and develop a variety of place-specific, small-scale creative tourism initiatives in small cities and rural areas throughout the Norte, Centro, Alentejo, and Algarve regions, which together comprise most of mainland Portugal (see figure 1.1).

What Is Creative Tourism?

Since the infamous definition of creative tourism put forth by Richards and Raymond (2000)—"tourism that offers visitors the opportunity to develop their creative potential through active participation in learning courses and experiences that are characteristics of the holiday destination where they are passed" (18)—an array of other definitions have followed (see, e.g., UNESCO 2006; Jelinčić and Žuvela 2012; Blapp 2015), offering different points of emphasis and stemming from different cultural and geographic contexts. This evolving attention to creative tourism, as Richards (2011) highlights, has been contextualized by a "creative turn" in tourism (and tourism studies) and has propelled the development of more flexible and authentic experiences involving processes of co-creation between host and tourist. It is important to note that this trajectory has been shadowed by the risk and potential dangers resulting from the commodification of everyday life through such tourism, which continues as an active point of concern in creative tourism development (see chapters by de la Barre and Jeannotte in this volume).

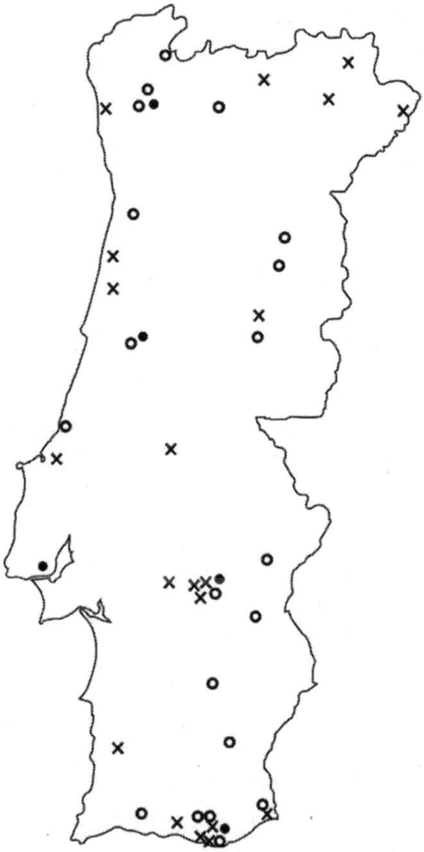

Figure 1.1. Map of Mainland Portugal Showing CREATOUR Pilots and Research Centres

LEGEND TO MAP:

o   1st call CREATOUR pilots
x   2nd call CREATOUR pilots
•   CREATOUR research centres

To allow for flexibility while also setting out some "working boundaries" around the question "What is creative tourism?" the CREATOUR project established the following definition of creative tourism: a sustainable, small-scale tourism that provides a genuine visitor experience by combining an immersion in local culture with a learning and creative process. To distinguish "creative" tourism from "experience" tourism, particular emphasis is placed on the creation process and capacity for the visitor to engage in the activity not only from the perspective of learning and skill development (or, alternatively, of entertainment and relaxation), but also of the potential for self-expression (Duxbury, Kastenholz, and Cunha 2019).

The vision of creative tourism guiding the project's pilot activities is centred on active creative activity encouraging personal self-expression and interaction between visitors and local residents, inspired by local endogenous resources (place and people), and designed and implemented by local residents (Duxbury, Silva and Castro 2019). In brief, the CREATOUR perspective on creative tourism includes four dimensions: active participation, creative self-expression, learning, and community engagement.

What Is CREATOUR?

Launched in November 2016, CREATOUR (Creative Tourism Destination Development in Small Cities and Rural Areas)[1] is a national 3.5-year research-and-application project to develop and pilot an integrated approach to creative tourism in small cities and rural areas in Portugal. The project aims to link interdisciplinary social science research with entrepreneurial and community-engaged practices in creative tourism. On the research side, its goal is to examine and reflect on creative tourism activities, including development dynamics and patterns, reception experiences, and community impacts, using methodologies and theoretical perspectives from the fields of tourism, cultural development, and local/regional development. On the practice side, it aims to catalyze creative tourism activities and providers in small cities and rural areas in Portugal, inform and learn from their development, and link them with each other through the development of a national network. This network-in-formation aims to offer not only visibility through critical mass, but also support through research, co-learning, and capacity building. In time, the project also aims to inform policy development relating to creative tourism.

The project was developed and funded in accordance with strategic public policy priorities to bridge the culture/creative and tourism sectors and to diversify tourism offerings. The CREATOUR project focuses on creative tourism as this bridge and as a strategic area for diversification. Although some existing (isolated) creative tourism workshops and related activities previously existed in the country, they worked in isolation from each other. While some of these are now pilots within CREATOUR, most pilot projects are new initiatives launched in response to the emergence of CREATOUR.

A key decision in the design of the project was to focus on areas outside Lisbon and Porto. In contrast to the overtourism experienced by many large cities and coastal resort communities, CREATOUR aims to develop attractive offers in other less visited areas, such as rural areas and small towns and cities. Furthermore, CREATOUR was inspired by a desire to create a structure that would provide value and a higher profile (both locally and nationally) to the many artistic and cultural organizations working outside the two big cities. To avoid potential "urban colonization of countryside" approaches, pilot project designers/providers must be based in these communities, not in Porto or Lisbon.

The project's design also responds to OECD advice that creative-sector development can be enhanced by policy measures and programs designed to build knowledge and capacity, support content development, link creativity to place, and strengthen network and cluster formation (OECD 2014). These dimensions form the framework for CREATOUR's approach. Linking creative activity to place in a context of tourists seeking "authentic" cultural and creative tourism experiences foregrounds the development of workshops and related participatory creative activities that are based on, and informed by, local history, traditions, and cultural expressions—and that are also envisioned, designed, and embedded locally.

Structure

CREATOUR is operationalized through research teams located at five research centres and forty pilot organizations across its four regions, and is advised by a small Advisory Council consisting of international experts and key national organizations (see figure 1.2).

Figure 1.2. Structure of CREATOUR

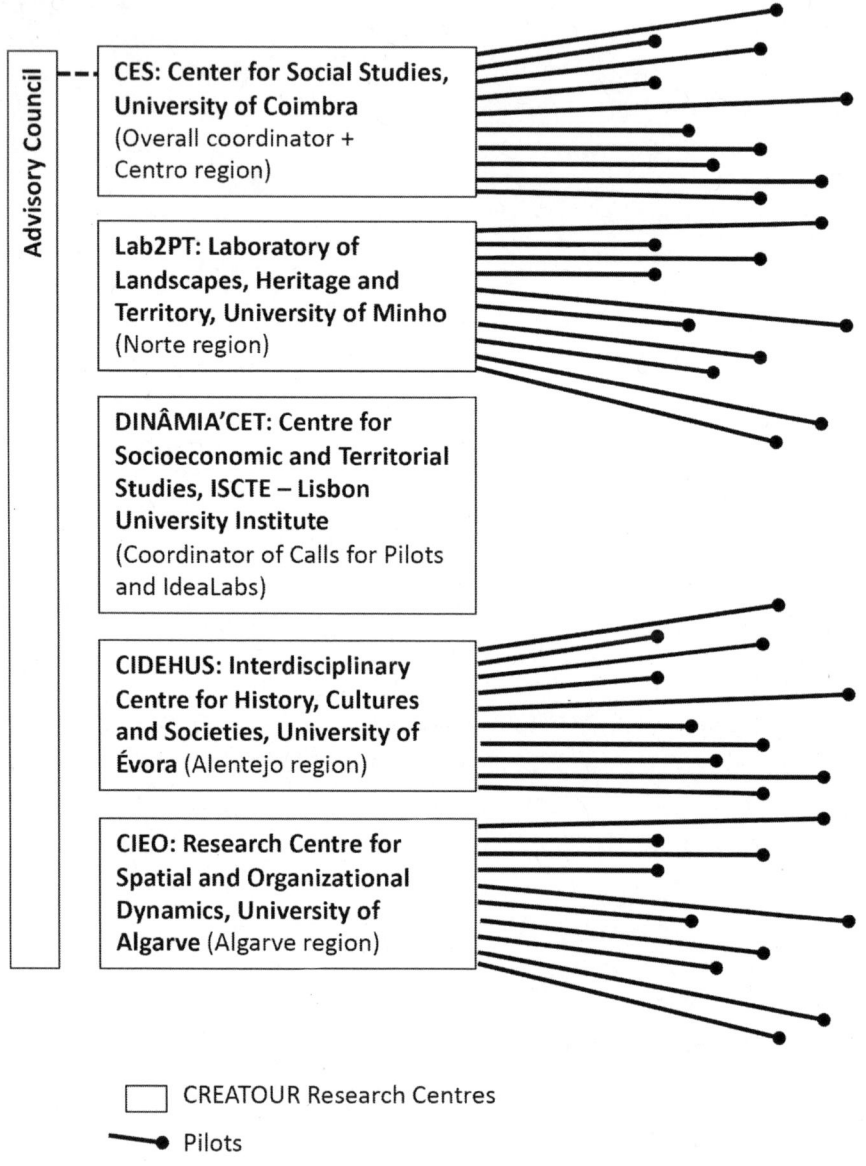

Advisory Council

CES: Center for Social Studies, University of Coimbra (Overall coordinator + Centro region)

Lab2PT: Laboratory of Landscapes, Heritage and Territory, University of Minho (Norte region)

DINÂMIA'CET: Centre for Socioeconomic and Territorial Studies, ISCTE – Lisbon University Institute (Coordinator of Calls for Pilots and IdeaLabs)

CIDEHUS: Interdisciplinary Centre for History, Cultures and Societies, University of Évora (Alentejo region)

CIEO: Research Centre for Spatial and Organizational Dynamics, University of Algarve (Algarve region)

☐ CREATOUR Research Centres

⟍● Pilots

Research Centres

The overall project is coordinated by the Centre of Social Studies (Centro de Estudos Sociais, or CES) at the University of Coimbra, which also serves as the in-region research centre for the Centro region. CES leads research on the development of an interdisciplinary theoretical framework for CREATOUR, coordinates the project's cultural mapping components, liaisons with the Advisory Council, provides administration and communications with funding bodies, and handles international relations. It also advises the other research centres on all aspects of the project. The Centre for Socioeconomic and Territorial Studies (Centro de Estudos sobre a Mudança Socioecónomica e o Território, or DINÂMIA'CET-ISCTE) at ISCTE–Instituto Universitário de Lisboa coordinates the national calls for pilots and the regional and national IdeaLabs, and conducts research at a national level on strategic actor-network analysis and the local impacts of the pilot initiatives. The Laboratory of Landscapes, Heritage and Territory (Laboratório de Paisagens, Património e Território, or Lab2PT) at the University of Minho is the in-region research centre for the Norte region and coordinates the development of contextualizing knowledge of tourism in Portugal and research on creative tourism platforms internationally. It is also leading the development of an augmented reality smart phone application that can be used to add a leading-edge digital dimension to creative tourism knowledge systems and experiences. The Research Centre for Spatial and Organizational Dynamics (Centro de Investigação sobre Espaço e Organizações, or CIEO) at the University of Algarve is the in-region research centre for the Algarve region and coordinates the evaluations of the IdeaLabs and the assessment frameworks for the pilot initiatives. The Interdisciplinary Centre for History, Cultures and Societies (Centro Interdisciplinar de História, Culturas e Sociedades, or CIDEHUS) at the University of Évora is the in-region research centre for the Alentejo region and coordinates the development of sustainability options and models for the continuance of CREATOUR beyond its initial 3.5-year period.

Pilots

The forty pilots within CREATOUR, selected in 2017 and 2018 (the process is described later in this chapter), develop and implement creative tourism activities and provide "front-line" knowledge and insights as co-researchers in the project. Ten pilots are located in each of the project's four regions and are situated diversely within each region (see figure 1.1 above). The pilot organizations include not-for-profit art and cultural associations, small entrepreneurial businesses, municipalities, regional development associations, and a few inter-organizational partnerships developed for the call. Appendix A below presents each of the pilots and a brief description of their projects. All pilot projects offer creative tourism activities that "embody local traditions or expertise, local history, and ways of life blended with other specifics of the cultural and local landscape of the locations where activities are organized, whether in a small city or a rural area. This place-sensitive development process served to connect creativity to place and create genuine, immersive, and creative experiences" (Duxbury, Silva, and Castro 2019, 297).

Advisory Council

The Advisory Council was established to provide external advice to the project in both its research and its application dimensions. Its international members are Dr. Greg Richards, NHTV Breda University of Applied Sciences and University of Tilburg, Netherlands; Dr. Patrick Brouder, Vancouver Island University, Canada; Caroline Couret, executive director of the Creative Tourism Network, based in Barcelona, Spain; and Marie-Andrée Delisle, an international expert on tourism marketing and communication, based in Montreal, Canada. Nationally, it includes the Intermunicipal Community of Central Alentejo (Comunidade Intermunicipal do Alentejo Central, or CIMAC), an intermunicipal bridging association that has participated in European projects incubating creative organizations in smaller-city contexts and has developed a creative ecosystem mapping initiative in its region, and ADDICT: Creative Industries Agency Portugal, an umbrella organization for the cultural and creative industry sector in Portugal.

Key Components

Three interlinking dimensions are core to CREATOUR: the pilot initiatives, the IdeaLabs, and the effort to strengthen network/cluster formation. These dimensions are informed and guided by complementary multidisciplinary research activities, knowledge-advancing annual conferences, and the development of a range of publications and other outputs.

Partner organizations, selected within the course of the project, develop and implement an array of *pilot initiatives* (i.e., creative tourism offers) in each year of the project. Monitoring and assessments of the pilot initiatives provide ongoing analysis of their processes, outcomes, issues, and impacts. CREATOUR gathers empirical data through activity data forms filled by organizations, visitor-participant questionnaires, site visits (and participant-observer field notes), "journey log" contents written by the pilots, interviews, and discussions at IdeaLabs and conferences.

IdeaLabs focus on providing support for content development and the linking of creativity to place. They provide regular points of contact to guide the development of pilot initiatives, support other project actions (such as cultural mapping, post-activities reflections and evaluations, and post-project sustainability discussions), and foster intra- and interregional organizational connections among cultural/creative organizations and with the tourism sector. Each year, two regional IdeaLabs take place in each project region, with an annual national IdeaLab taking place with each of the annual conferences.

The project's focus on "strengthening network" and "cluster formation" focuses, first, on strategies to facilitate linkages between the pilots and, second, on the development of post-project sustainability options and strategies. This work focuses attention on the broader relations among the organizations presenting creative tourism experiences, models, and strategies for cross-sector alliances with tourism and other community actors, and roles for local authorities and regional agencies. It is informed by, and developed through, interactions and discussions in IdeaLabs, cross-sectoral focus groups and meetings, consultations with the project's Advisory Council, and international best practices.

On the research side, in addition to monitoring the pilot initiatives, CREATOUR is developing a system for monitoring and creating a baseline of knowledge to track macro-changes during the project and place project activities and findings in wider contexts. This includes an international scan and analysis of best practices and issues in creative tourism projects and networks in other countries, monitoring shifts in tourism flows in each region (in conjunction with regional tourism agencies and observatories), and investigating the state of knowledge and characteristics of cultural/creative organizations located outside the Lisbon and Porto Metropolitan Areas.

Since much learning occurs through encounters and discussions, and because we wish to both inform ourselves of international practices and share with others what we are doing, CREATOUR organizes an annual international conference on creative tourism, held in a different region each year (Centro, Norte, and Algarve). The 2019 conference also featured a creative tourism showcase. An international conference linking cultural mapping and intangible cultural heritage was organized in the Alentejo region in 2018. The project also develops three types of publications: academic, professional, and policy-oriented. In addition to academic works, CREATOUR has developed a publication directed at the cultural/creative and tourism sectors focusing on good practices, lessons learned, and recommendations for action (Vinagre de Castro et al. 2020). A briefing report with policy recommendations is directed to local authorities and regional agencies (culture, tourism, regional development, etc.) (Gonçalves et al. 2020). The project also informs the development of an online course (MA level), summer schools, and a video documentary.

Reflections on CREATOUR's Operationalization

In traversing the distance from "written proposal" to "implementation and forward movement," adapting to the realities emerging from such a multi-dimensional research-and-application project has been challenging but also vital and interesting. The project is both guided and restrained by what was anticipated in the initial proposal. While some aspects have been revised in a minor way to reflect the implementation conditions, the addition of unanticipated actions is constrained by the approved budgets and plans. Moreover, the unexpected has provided important learning moments and opportunities and has led to the sharing of diverse perspectives that have sparked new ways of understanding and interpreting developments. We discovered that ambitious work plans, the complexities of launching "new" tasks and coordinating geographically distributed teams, and dealing with rapidly advancing timelines and expectations of "support" have been challenging but also important dimensions in the evolving learning and development process we have undertaken.

Selecting Pilots

CREATOUR's forty pilot projects were selected through two national open calls for which all types of organizations were eligible, with twenty pilots selected in early 2017 and twenty selected in early 2018. In each call, five organizations were selected in each region: Norte, Centro, Alentejo, and Algarve. Applications were reviewed according to the following criteria:

- cultural value of the activities proposed;
- the creative nature of the activities proposed;
- capacity of tourism attractiveness;
- impact of the project in terms of community development;
- diversity of focus of the proposals;
- and capacity and commitment to work with the research team during the project.

In each year, the selection of the pilots occurred in a two-phase process: first, the in-region research centre reviewed applications from their region and pre-selected approximately seven candidates. Second, all five research centres met to learn about and review the pre-selected candidates, and to reflect on and discuss the "national" picture that resulted. Shortly after this meeting, each in-region team finalized their selection of five pilots; the full set of twenty pilots was then announced. Overall, the pilots were selected so as to include a wide diversity of approaches, activities, organizational types, and geographic coverage. In the second call, pilot-applicants were also assessed in terms of how they complemented the range of pilots selected in the first call and strengthened the overall network. The selected pilots signed contracts with their in-region research centre, committing to defined research-related tasks during the time of the project in return for a fee acknowledging the time and efforts—and knowledge—they will be contributing to the project. The pilots play a central role in the project through the development and implementation of a set of creative tourism activities and are viewed as co-researchers in CREATOUR.

Fostering Relations

Structural dimensions to foster pilot-researcher interaction and knowledge exchange/co-learning include in-region research teams, regional IdeaLabs, and an annual national IdeaLab and international conference. Site visits during selected creative tourism offers, which involve participant observation, documentation, and in situ interviews with the pilot organizer are also planned annually with all pilots. Additional events organized by the in-region research centres, a closed Facebook group, and an email listserv have provided additional points of interaction and knowledge sharing. Nonetheless, there remains a feeling of disconnection when the pilots are in their home communities preparing their activities, and when network dynamics and the "feeling" of operating in a network context are still in formation (for more information on strategies used to foster research-practice collaboration within CREATOUR, see Duxbury, Bakas, and Carvalho 2019).

Anticipations and Realities among the Pilots

Among the many things learned during the project, there have been a few "surprises" that provided opportunities to reflect on the initial project design as written in the funding application. These have resulted in valuable insights into the pragmatic realities of developing creative tourism offers in non-metropolitan contexts.

The first surprise was the nature of the pilot organizations—that is, the applicants to the CREATOUR project and the final selection of pilot organizations. The initial expectation was that the pilots would be primarily arts organizations and cultural associations for whom creative tourism could become an additional revenue stream to complement their "core" artistic/cultural work. Secondarily, a number of cultural centres housed in renovated former factories were expected, with creative tourism providing an additional activity in these renovated spaces and a way to enhance the connections between local artists and creative enterprises and visitors to their area. While each of these "types" is found among the CREATOUR pilots, they are not in the majority. Among the initial twenty pilots, a variety of organization types were selected, including not-for-profit art and cultural associations, but also small entrepreneurial businesses, municipalities, regional development associations, and a few multi-organizational partnerships developed for the call. Many of the entrepreneurs are younger people who are looking to launch a creative tourism business, often in conjunction with other work and activities they currently organize and conduct.

Leading roles have been taken up by municipalities, regional development associations, and a few private entrepreneurs (see Bakas, Duxbury, and Castro 2018) to launch and coordinate local networks of creative tourism offers in collaboration with a range of independent individuals or organizations. These regional or local networks of activities aim to involve a range of groups that collaborate to offer events and activities. In this way, the pilot agent is catalyzing locally an array of activities to be launched in a collaborative manner, by a range of different local actors. Other organizational models included the integration of creative activities as the defining feature of small-scale festivals and tradition-based businesses that include creative tourism and other types of activities, with attention to both business development as well as the wider socio-cultural and economic development of the community in which they are based. For most pilots, "creative tourism" is a new addition to a portfolio of other activities.

A second type of surprise was the nature of the creative tourism offers. While the project proposal envisioned an initial development of simple workshops that would grow and be augmented over time, instead many pilot proposals entail the development of much more comprehensive offers and enterprises, as mentioned above. This discrepancy may stem, in part, from the fact that many of the prominent promoters of creative tourism in its "earlier" years have been large cities. While creative tourism initiatives can increasingly be found in smaller places and rural areas, an implicit "urban context" may have influenced research and practice in the field in ways we may not yet fully recognize. For example, in a large urban context like Barcelona or Paris, a photographer can offer a simple workshop on a weekend without the need for a broader strategic business plan relating to creative tourism. However, in a smaller, more remote place, much more attention must be directed to, for instance, niche marketing and attraction; the scale of activities to warrant the travel distance; the specificities of a "place" that nurture and inform the activities and provide distinctiveness to the creative tourism offers; issues related to developing an enterprise to sustain the activities (and to balance with other initiatives); and even pragmatic issues of transport become heightened. The extent of the support and mentoring that might be needed in regards to such business start-up issues was not fully realized in the initial application. The necessity of networking among creative tourism entities in smaller places is emerging as key not only for visibility, but also for knowledge sharing and capacity development. Linking this networking to addressing such start-up issues is an identified challenge on our agenda moving forward.

Within this expanded context, the nature of the creative tourism initiatives that were proposed and are in the course of being implemented range widely. Workshops relating to traditional arts and crafts, which often preserve and pass on skills and knowledge between elders and younger generations, are popular. For example, the linen cycle (from seeding to weaving) and wool processing (from preparing to natural dying to weaving) have inspired a series of activities. Contemporary artistic expression is also the basis of some pilot activities—for example, workshops and other initiatives related to advancing one's photography skills in the unique landscapes of the Serra Estrela or the dark sky areas of the Alentejo, or to photographing the people and settings of traditional fishing-related activities in Nazaré or other traditional activities in villages within the Peneda-Gerês National Park. There are marble sculpting workshops—where attendees can create their own sculptures—planned for the Rota do Mármore do Anticlinal de Estremoz region and in Évora. The Roman mosaic heritage of Conímbriga, the Roman villa of Rabaçal, and the Monumental Complex of Santiago da Guarda have inspired a series of events—including mosaic-building workshops—to learn about the region's Roman mosaic heritage and to reinterpret it for both the present and the future.

Gastronomy classes focusing on traditional regional dishes are also featured among the pilot projects. Notably, within a creative tourism context the focus is on food preparation and learning (and cooking as a creative activity) rather than on just "trying and tasting" new foods;

examples range from "how to cook cataplanas" in the Algarve to "traditional couscous" cooking in Bragança to the creation of traditional sweets with regional products. Related to this, there is interest among a number of pilots in sharing knowledge about wild plant resources in various locales and revitalizing activities and practices related to their traditional uses. These represent "additions" to the initial expected focus on arts and crafts, reflecting the widened definition of "culture" that must be adopted in smaller communities and rural areas in order for the locally important aspects of place to resonate within the creative tourism context.

Emphasizing the Specificities of Place: Autonomy . . . with Support

As the project has proceeded, CREATOUR has placed a strong emphasis on place and local initiative. These have been operationalized on a number of levels, and as Duxbury, Silva, and Castro (2019) write, "local residents lead the design and development of the pilot activities, local community involvement in the activities is strongly encouraged, and significant and 'special' natural and historic places provide some contextualizing sites for the activities" (297). We have encouraged a variety of approaches and initiatives among the pilots selected, and have encouraged them to highlight the specificity of particular locales. Part of the research entails monitoring and exploring how a wide range of activities develop, to track different development patterns and learn from the various strategies and projects implemented.

However, local autonomous action requires a support system that, as we have learned, must be more extensive and ongoing than the actions initially envisioned within CREATOUR. In the early start-up phase, a number of issues emerged prominently in discussions with the pilot organizations. Communication and promotion issues linked to uncertainty about target audiences and how to reach them, which is in turn related to the production-led (or supply-led) model adopted within CREATOUR, with activities primarily based on local resources and pilots' interests and skills. As discussed in Duxbury, Silva, and Castro (2019),

> We contend that this is an appropriate guiding strategy in the context of a creative experiment and the desire to foster a new

and different outlook, to diversify current offers, to surprise and to be flexible and innovative in development trajectories. In this context, encouraging grassroots-led experimentation and new ideas is the project's initial priority, with importance of "market" becoming secondary once the initial ideas and offers are defined. However, this creative ideas-based approach also brings a risk (*cf.* Raymond, 2007) concerning the degree of "take up" of the offers, and also the speed at which this could happen: on the one hand, there is the nagging question, "Will anyone be interested in what we want to offer?" and, on the other hand, there are pragmatic questions about the operational capacity of each pilot and the carrying capacity of the small communities in which they are embedded. (299–300)

The development process has been complicated by the reality that creative tourism is a side activity for the pilot organizations, so limited time can be devoted to these new initiatives and sustainable business models are not readily available for referencing, so the question of making activities economically sustainable in the medium-term is an ongoing concern—and a topic for interlinking research and practice.

Also, we learned that it is essential to plan for the time necessary for start-up activities to be carefully designed, planned, and implemented. This process is lengthened by the involvement of local community organizations and multiple actors. Building a network also requires time for meeting, interacting, learning together, building trust, and collectively deciding on and taking action. Furthermore, we have come to realize that "although the small scale and flexibility of the creative tourism offers encourages learning-by-doing and processes of incremental improvement, and the IdeaLabs offer initial guidance and processes through which suggestions can be shared and ideas developed further, the establishment of an 'R&D' phase for both product development and audience research—before pilots are formally offering tourism products to the public—may be advisable within similar projects" (Duxbury, Silva and Castro 2019, 301).

Establishing Research Platforms and Processes

CREATOUR is an interdisciplinary project approaching creative tourism from a variety of angles, aiming to intertwine tourism, culture, and local/regional development perspectives. Each of the research teams brings different experiences and areas of expertise so the project is catalyzing interactions across geographic regions as well as disciplines and building relations among researchers who had not previously worked with one another. This has added an additional layer of complexity and challenge, and we are continually learning and striving to develop good practices in communication, transparency, and collaboration while also catalyzing a network among the pilots, and between the researchers and pilot-practitioners in each region.

Anticipations

In non-chemistry contexts, *catalyze* is a transitive verb meaning "bring about" or "inspire" (*Merriam-Webster*), or "to cause (an action or process) to begin" (*Oxford English Dictionary*). To date, the launch and development of CREATOUR has demonstrated the complexities involved in this catalyzing process, and the need for flexibility, dynamic relations, and attentiveness to the variety of perspectives brought together within this initiative.

CREATOUR is an experimental project to learn how creative tourism activities can be launched and become sustainable in small city and rural contexts, perhaps by being combined with other initiatives, and how they can serve as a new revenue source and bring other benefits (or issues) to the local organizers and host communities. By nature, the individual activities are small-scale, but a regular stream of such activities may provide a seed for wider development processes. As one of our pilot project organizers has stated, "A small number of people in a small place can have a large impact."

Creative tourism can provide a small-scale tourism option for smaller places. It attracts visitors who want to learn and engage with a place and its residents in a meaningful and creative way, and the attraction is activity-specific and interest-led (see Richards 2011). This enables locales that are "outside the regular tourism circuits" to attract visitors, to keep them for a period of time (to participate in the creative tourism workshops and related activities), and potentially to attract them outside the "usual high-tourism season."

Creative tourism is also one avenue of art/culture-based development, with cultural activities as a driver and enhancer of local sustainable development, which is a growing phenomenon and focus of interest in smaller communities internationally. Creative tourism can provide an additional revenue stream for local artists, artisans, and designers/creators, and a platform for local entrepreneurial energies, skills, and the building of collaborative local networks. It can serve to highlight local traditions and cultural assets and revitalize them in contemporary settings.

Creative tourism is a relatively young field of international research; for example, there is still little sustained research into the development processes of creative tourism initiatives, nor "sustainable creative tourism models" in operation, nor systematic research into the multi-dimensional impacts of these activities (cultural, social, economic, and other) in smaller communities. And there is little systematic knowledge about "who is the creative tourist" and what visitor profiles are attracted to different creative tourism activities. We hope to inform these gaps in knowledge. However, CREATOUR is more than a tourism research project; we also intend to extend knowledge of good practices as well as critical issues involved in enabling culture-based development initiatives in small cities and rural areas—an emerging field internationally—and in linking culture and tourism organizations in these settings.

The challenges faced by culture-based and creative organizations/businesses based in small communities, rural areas, and sometimes remote settings are substantial matters to address in designing appropriate research, networking, support, and policy structures (Nelson, Duxbury, and Murray 2012; Luckman 2012; Collins and Cunningham 2017). Substantial support for the start-up phase—especially in the design and planning processes and early experimentation activities—appears to be necessary, but may not be required in later operations (depending on the nature of the activities). CREATOUR pilots are encouraged to think about multi-dimensional community benefits relating to their activities, and to build this into the core of their business planning (rather than merely consider it as an "add-on" later on). These community-focused and -engaged dimensions may warrant and benefit from public support to enable broader community benefits to unfold and new initiatives to germinate. As Brouder (2019) argues, the development of community-focused networks and "a focus on community relationships, rather than tourism relationships per se," fosters an environment of support for creative tourism. Facilitating and nurturing such virtuous cycles could have considerable positive effects on the long-term sustainable development of these communities.

NOTES

This chapter was developed within the CREATOUR project (no. 16437), which is funded by the Portuguese Foundation for Science and Technology (FCT/MEC) through national funds and co-funded by FEDER through the Joint Activities Programme of COMPETE 2020 and the Regional Operational Programmes of Lisbon and Algarve.

1   The CREATOUR website is available at www.creatour.pt.

2   ADRAT is the owner of the Barro Negro de Vilar de Nantes brand, recognized by the National Industrial Property Institute, Portugal, number 212/2011 (11 July 2011).

References

Bakas, Fiona Eva, Nancy Duxbury, and Tiago Vinagre de Castro. 2018. "Creative Tourism: Catalysing Artisan Entrepreneur Networks in Rural Portugal." *International Journal of Entrepreneurial Behaviour and Research* 24, no. 1. Special issue on Artisan, Cultural, and Tourism Entrepreneurship. https://doi.org/10.1108/IJEBR-03-2018-0177.

Blapp, Manuela. 2015. "Creative Tourism in Bali's Rural Communities: Examination of the Current Offer and Advice on Future Product Development." MA diss., NHTV University of Applied Sciences, Breda. http://www.tourism-master.com/wp-content/uploads/2016/03/Master-Thesis-Manuela-Blapp-10.12.15-electronic-version.pdf.

Brouder, Patrick. 2019. "Creative Tourism in Creative Outposts." In *A Research Agenda for Creative Tourism*, edited by Nancy Duxbury and Greg Richards, 57–68. London: Edward Elgar Publishing.

Collins, Patrick, and James A. Cunningham. 2017. *Creative Economies in Peripheral Regions*. London: Palgrave Macmillan.

Duxbury, Nancy, Fiona Eva Bakas, and Claudia Pato de Carvalho. 2019. "Why Is Research–Practice Collaboration So challenging to Achieve? A Creative Tourism Experiment." *Tourism Geographies*. Special issue on Creative and Disruptive Methodologies in Tourism Studies. https://doi.org/10.1080/14616688.2019.1630670.

Duxbury, Nancy, Elisabeth Kastenholz, and Conceição Cunha. 2019. "Co-producing Cultural Heritage Experiences through Creative Tourism." In *E-CuL-Tours: Enhancing Networks in Heritage Tourism*, edited by Werner Gronau, Rossana Bonadei, Elizabeth Kastenholz, and Albina Pashkevich, 189–206. Rome: tab edizioni.

Duxbury, Nancy, and Greg Richards. 2019. *A Research Agenda for Creative Tourism*. Cheltenham, UK: Edward Elgar Publishing.

Duxbury, Nancy, Sílvia Silva, and Tiago Vinagre de Castro. 2019. "Creative Tourism Development in Small Cities and Rural Areas in Portugal: Insights from Start-up Activities." In *Creating and Managing Experiences in Cultural Tourism*, edited by Daniella Angelina Jelinčić and Yoel Mansfeld, 291–304. Singapore: World Scientific Publishing.

Gonçalves, Alexandra R., Maria do Rosário Borges, Nancy Duxbury, Cláudia Pato Carvalho, and Pedro Costa. 2020. *Policy Recommendations on Creative Tourism Development in Small Cities and Rural Areas*. Coimbra: CREATOUR project, Centre for Social Studies, University of Coimbra. https://creatour.pt/en/publications/policy-recommendations-on-creative-tourism-development-in-small-cities-and-rural-areas/.

Jelinčić, Daniella Angelina, and Ana Žuvela. 2012. "Facing the Challenge? Creative Tourism in Croatia." *Journal of Tourism Consumption and Practice* 4, no. 2: 78–90.

Luckman, Susan. 2012. *Locating Cultural Work: The Politics and Poetics of Rural, Regional and Remote Creativity*. London: Palgrave Macmillan.

Nelson, Ross, Nancy Duxbury, and Catherine Murray. 2012. "Cultural and Creative Economy Strategies for Community Transformation: Four Approaches." In *The Social Transformation of Rural Canada: New Insights into Community, Culture and Citizenship*, edited by John Parkins and Maureen Reed, 368–86. Vancouver: UBC Press.

Organisation for Economic Co-operation and Development (OECD). 2014. *Tourism and the Creative Economy*. OECD Studies on Tourism. Paris: OECD Publishing.

Raymond, Crispin. 2007. "Creative Tourism New Zealand: The Practical Challenges of Developing Creative Tourism." In *Tourism, Creativity and Development*, edited by Greg Richards and Julie Wilson, 146–57. London: Routledge.

Richards, Greg. 2011. "Creativity and Tourism: The State of the Art." *Annals of Tourism Research* 38, no. 4: 1225–53.

Richards, Greg, and Crispin Raymond. 2000. "Creative Tourism." *ATLAS News* 23: 16–20.

UNESCO. 2006. *Tourism, Culture and Sustainable Development*. Paris: UNESCO.

Vinagre de Castro, Tiago (Coord.), Nancy Duxbury, Sílvia Silva, Fiona Bakas, Cláudia Carvalho, Lorena Sancho Querol, Rosário Borges, Sara Albino, Alexandra Gonçalves, Paula Remoaldo, and Olga Matos. 2020. *Creative Tourism: Guide for Practitioners*. Coimbra: CREATOUR project, Centre for Social Studies, University of Coimbra. https://creatour.pt/en/publications/creative-tourism-guide-for-practitioners/.

APPENDIX A

CREATOUR PILOTS AND PROJECTS

Forty CREATOUR pilots, located in four regions of Portugal, are developing a wide range of creative tourism offers, inspired by and embedded within the cultures and locales in which they are operating. All activities are designed and implemented locally. This appendix presents a brief overview of the pilots, organized by region (north to south), with the title and a brief description of their initial pilot projects. As an experimental learning and development experience, the pilots are encouraged to evolve, adapt, and develop new offers as they go forward.

NORTE

Associação de Desenvolvimento da Região do Alto Tâmega (ADRAT) (regional development association), Revitalizar Vilar—Revitalização da Olaria Negra de Vilar de Nantes (Revitalize Vilar—Revitalizing Vilar de Nantes Black Pottery). This project intends to preserve and promote the local pottery handicraft brand Barro Negro de Vilar de Nantes[2] by promoting local and regional socio-economic activity and fostering a process of sustainable development through contact between visitors and the population of Vilar de Nantes and their culture. Local potter-artisans skilled in the region's traditional pottery techniques organize and hold workshops on black pottery (a unique type of pottery from this area) in which participants learn about and participate in the whole production cycle, from collecting raw clay to baking the final clay pots and other objects.

Associação de Desenvolvimento das Regiões do Parque Nacional da Peneda-Gerês (ADERE-PG) is a non-profit development entity with interventions in the five municipalities of the Peneda-Gerês National Park. The Creative Experiences with Sense(s) project involves activities in the five municipalities that are part of the park, based on the five senses and connected to local traditions: "Art in Transhumance" (touch) in Castro Laboreiro, Melgaço; "Discovering Folklore" (hearing) in Lavradas, Ponte da Barca; "Honey Secrets" (taste) in São Jorge, Arcos de Valdevez; "Natural Pantry" (smell) in Covide, Terras de Bouro; and

"Ethnographic Photography: Linen Heritage and Memories" (sight) in Cabril, Montalegre.

Câmara Municipal de Bragança (municipal government), Cá se Fazem Cuscos: Workshop de Confeção de Couscous de Bragança (Here Couscous Is Made: Homemade Couscous of Bragança Workshop). This weekend activity demonstrates the cycle of production and traditional cooking of couscous, a food particular to this area within the Portugal gastronomic landscape. The municipality organizes the overall project, with local artisans holding tours and implementing the production/preparation workshops.

Desteque—Associação de Desenvolvimento da Terra Quente Transmontana (development association), Pelo Fio do Fato se Conhece o Careto: Oficinas de Tecelagem e Latoaria (Unmasking the Careto through the Thread of the Costume: Weaving and Tin Workshops). Based on the UNESCO-designated intangible heritage of the area, these workshops have been developed to make the masks, cloaks, and animal bells that carnival figures (*Caretos*) use during the Carnaval festival. The project is organized by the association in partnership with local artisans, and is held before the Carnaval.

Galandum Galundaina (cultural association), Festival "L Burro i L Gueiteiro" ("Donkey and Bagpiper" Festival). This itinerant festival of traditional Mirandese culture, which travels from village to village in the region, is based on two important vectors for the local cultural heritage: the Mirandese donkey and the Mirandese bagpipe. During the afternoons of this festival, there are workshops on themes as diverse as the Mirandese language, the construction of traditional instruments, bagpipe and pastoral flute, "donkey knowledge" about the Asinine breed, the traditional *Pauliteiros* dance, and traditional percussion. In the evenings, concerts are held featuring lively music and dancing. The festival is organized in partnership with AEPGA (Associação para o Estudo e Proteção do Gado Asinino) and PALOMBAR (Associação de Conservação da Natureza e do Património Rural).

LRB—Investimentos e Consultoria LDA, Município de Boticas, Município de Guimarães, and **Município de Montalegre** (a partnership between a technology company and three municipal governments), Creative Tourism in the Territories of Montalegre, Boticas and Guimarães in Augmented Reality. Through this project, several themed creative experiences incorporating augmented reality are being developed for tourists, building on the natural and traditional cultural resources of each location. For example, workshops focusing on the cycle of wool production in the Barroso region have been developed, highlighting topics such as pastoralism, village life, and community work.

The **Municipality of Amares** has offered, biannually since 2009, Encontrarte Amares, a pluridisciplinary festival of visual arts and animated cinema bringing together the radicality of contemporary artistic thought with the traditional heritage of the region of Minho. The CREATOUR pilot, ARA—the Artistic Residencies Amares Co-creation Project, offers visitors an experience of active participation in artistic creation. For six days, national and international artists, the local community, and visitors—inserted into a very peculiar context—share, collaborate, and create artistic interventions rooted in the Amarense heritage, forming a co-operative moment of reflection and creation, gastronomy, art, and heritage.

Município de Esposende (municipal government), Escriativo: The Art of Reed. The core of Esposende's creative tourism initiative comprises traditional workshops in which participants make colourful woven baskets from local reeds. The municipality organizes and promotes the project in which local artisans teach. For example, within World Tourism Day celebrations, the "Basket of Reed" creative workshop allowed participants to learn two parts of the reed-basket-making process: work on the loom and sewing the pieces. The "Reed Goes to the Beach" workshop at Suave Mar beach, Esposende, took advantage of the presence of many bathers to develop a creative workshop and hold a fashion show, with articles of reed displayed in the sandy environment.

Turismo Industrial de S. João da Madeira—Município de S. João da Madeira (municipal government), Creative Industrial Tourism. Bringing together an established program of "industrial tourism" with "creative tourism" activities, the municipality organizes and promotes a variety of workshops in an industrial context within municipal buildings/factories, where visitors can learn industry-related techniques from former factory employees in areas such as shoe-making, hat construction, pencil-making, and making cookies in a commercial bakery.

The small enterprise **VERde NOVO** focuses on the cycle of linen production as one of the ancient traditional activities in two small villages, Cerva and Limões, which are set in a very peculiar and unique rural landscape that also serves as an inspiration for this endeavour. In the CREATOUR Linho de Cerva e Limões: Tecendo o Futuro pilot project (Linen Craft from Cerva and Limões: Weaving the Future), visitors are guided to participate actively in several steps related to the linen cycle, from sowing, harvesting, and dying, to weaving. The workshops and other activities involve local weavers who work in the communities using traditional techniques. VERde NOVO is also involved in organizing events related to heritage, tourism, and sustainable development.

CENTRO

ADXTUR—Agência para o Desenvolvimento Turístico das Aldeias do Xisto (regional development association), Schist Villages. Pottery, wood sculpture, and miniature "schist houses" workshops are held at a traditional schist village, Cedreira Village, with the local owner/proprietor organizing and providing the workshops herself. Workshops are available on a ongoing basis to people who stay at the accomodation in Cedreira Village and to others outside the village. Activities are promoted by ADXTUR through its online platform BookInXisto. Artistic residencies and additional creative activities/workshops at other schist villages are planned.

Associação Destino Caldas (tourism association), Caldas Creative Tourism. This CREATOUR pilot project organizes three different thematic tours of the city related to the specific culture and history of Caldas, including workshops to make objects related to some of the following themes: paintings, ceramics/design, and local market-related objects. For example, an interactive tour of the historic centre of the city has been realized in a creative itinerant theatre experience featuring the sites most closely identified with the master artist Rafael Bordallo Pinheiro, culminating in a workshop of modelling and painting of tiles with the colours of Caldas da Rainha ceramics.

Associação Dominio Vale do Mondego (cultural association), Mondego Art Valley. This annual, week-long arts festival is held at a biodynamic farm in the Serra da Estrella region. Festival participants (artists, volunteers, and visitors) actively participate in creative workshops dealing with such topics as theatre, stand-up comedy, singing, dancing, wood sculpture, felting, mosaic-making, and graphic design/animation. Each evening there are artistic performances, some created within the workshops.

Associação Luzlinar, a cultural organization based in the village of Feital, uses the rocky and mountainous Beira Alta landscape as both inspiration and setting for contemporary art workshops (focusing on, for example, photography, video, sketching/painting, and music). In the pilot project Campus Jardim das Pedras (Garden of Rocks), Luzlinar develops creative tourism workshops to lead the preservation and revitalization of shepherds' shelters and routes in the region (in which visitors participate). In this sense, Luzlinar promotes culture-based, creative activities that will bring visitors to the local villages and also preserve and disseminate local heritage.

Spanning three municipalities—Condeixa-a-Nova, Penela, and Ansião— the **Mosaico—Conímbriga and Sicó** partnership involves an array of creative tourism activities based on the Roman mosaic heritage present in this territory, which is rich in materials, techniques, decorative motifs, images, and narratives. MosaicoLab, for example, is held at the Monographic Museum of Conímbriga, located near an archeological site comprising the ruins of a Roman spa town, with many of the building's floor mosaics still intact and available for viewing. Inspired and informed by the mosaics at this site, MosaicoLab offers mosaic workshops that allow participants to learn about materials and techniques and to design and make their own small mosaic to take home. MosaicoLab has also developed an integrated creative program with local schools, teaching students and training teachers to deepen the connection between the communities and their cultural heritage. Overall, there are three distinct archeological areas where workshops are held, and the involvement of the local youth is a central focus in the development of Mosaico activities.

Município de Abrantes and **Canal 180** (municipal government partnership with online cultural broadcaster), 180 Abrantes Creative Camp. This partnership, a week-long creative summer "boot camp" for young people, promotes and organizes a series of activities under the theme of "Creative Collaborations in Media Arts." Invited national and international artists provide an array of creative workshops for the participants. Creative Camp participants make art pieces that remain within public spaces (urban installations based on elements of the local identity).

Quico—Turismo, Lda. (tourism company), Nazaré Criativa (Creative Nazaré). Inspired by both past and present cultural expressions and ways of life in Nazaré, and working with local artisans and creators, a variety of creative activities have been organized. For example, a photography expedition, with a cultural-historic focus, visits the fishermens' area and the city's historic space. Another activity is a sewing workshop based on local traditions, adapting the shape of the *carapau* (dried horse mackerel) to make a fish-shaped key chain.

Ruralidades e Memórias (cultural association), Tourism 3B. Focusing on the recovery and continuation of three important local artisanal activities in the village of Covão do Lobo, a series of workshops have been designed based on traditional ancestral techniques of processing resin (*breu*) and reeds (*bunho*), along with basket-weaving and traditional housing construction (with clay and sand) (*barro*/adobe). Offered by local residents with this knowledge, the project aims to instill pride in these local traditions and to provide an avenue to inspire younger generations and international makers to learn these techniques.

Tecitex (New Hand Lab) and **Museu Lanifícios UBI** (partnership of cultural entrepreneur/association and wool museum operated by the University of Beira Interior), Coolwool: Creative Weekend at Covilhã.

This partnership aims to hold creative weekends (and smaller events) that will include an array of activities organized within the museum of wool and the artist ateliers situated within the old wool factory. Steeped in the historic and more contemporary industrial wool processing traditions of Covilhã, participants will have the opportunity to participate in a range of workshops, includings some involving interactive fabrics and historical textiles.

VIC // **Aveiro Arts House** (cultural association), VIC // Ossos do Ofício ("All Part of the Job"). The Aveiro Arts House, a former pottery and cinema artist's house and atelier, includes a guest house, mini-cinema, art studio, and art gallery with its own collection. It also hosts artists-in-residence who offer workshops in ceramics, serigraphy, kinetic scupltures, travel writing, filmmaking, and music. Guests and workshop participants have close contact with artists within the artistic environment that the house provides. The workshops are offered for people staying in the guest house as well as others in Aveiro (visitors or residents), offering creative points of interaction, inipiration, and possible co-creation.

ALENTEJO
CACO—Associação de Artesãos do Concelho de Odemira (cultural association), Mãos de Cá (Hands of Here). Located in the Alentejo Coast region of Portugal, this association of locally based artisans organizes and promotes workshops featuring both traditional and contemporary crafts. Through these activities, CACO aims to both attract new visitors to the municipality of Odemira and introduce sustainability and innovation in traditional arts and crafts. From its Arts and Crafts Centre, the organization launches creative activities that take place in the studios of its associates for the production and/or creation of products in weaving, sewing, jewellery, sculpture, pottery, and carpentry.

Câmara Municipal de Mértola (municipal government), Creative Walks. Within a renowned natural setting, a variety of thematic and interpretive nature-based walks are offered; the aim is to discover the resources within the local natural environment and participate in interconnected nature photography workshops and wild herb picking. They also organize astronomy observation in a UNESCO-certified dark sky reserve where there is no light pollution, during which professional photographers offer participants guidance and advice on night photography. During the Nights in the Market event, held at Mértola's traditional market, gastronomy workshops offer participants opportunities to get to know different aspects of Mediterranean cuisine.

Centro de Arte João Cutileiro, dynamized by the cultural association Pedra+, works on issues and activities relating to ornamental stone waste and access to culture. In the Pedra+ (Stone+) project for CREATOUR, the art centre offers artistic residencies and workshops for professionals and amateurs with some knowledge in the techniques of stonework. Visitors can use the stone-carving machines of the artist João Cutileiro in his own atelier and try stone drawing. Pedra+ also organizes visits with creative studio activities for the general public.

Centro de Estudos de Cultura, História, Artes e Patrimónios (CECHAP) (cultural and development association), Rota do Mármore do Estremoz Anticlinal (Marble Route of the Estremoz Anticlinal). Focusing on the "Zona dos Mármores" (Zone of Marbles) area in which it is located, CECHAP is an active player in safeguarding the cultural identities of the area's communities and awakening younger generations to their culture. Embedded within local marble traditions, CECHAP organizes workshops to teach visitors about how to work the stone. With the support of a master artisan who explains the different facets of the marble stone (e.g., its textures, colours, densities, etc.) and shows different ways of working the stone with diverse tools, visitors learn through experimentation and the use of traditional techniques. Reusing "waste" marble, visitors are challenged to create marble panels that they can take home.

Genuine Alentejo—Portugal's Remarkable Tourist Experiences (tourism animation agent), Genuine Alentejo. Operating within the Alentejo region, this company designs and organizes tourism products and programs based on "hands-on" experiences linked to some of the following historical themes: heritage and culture; wines and gastronomy; arts and crafts; nature, landscapes, and sports; and contemporary performing arts. Within CREATOUR, Genuine Alentejo collaborates with various organizations offering creative tourism workshops and other activities, bringing them within broader tourism programs. By integrating visitors into the identity and true essence of local communities and facilitating opportunities to participate in creative learning experiences, Genuine Alentejo aims to contribute to the sustainable development of the region while respecting and helping to preserve its ethnographic, natural, historical, and cultural heritage.

MARCA—Associação de Desenvolvimento Local (local development association), Saídas de Mestre (Master Outings). Operating in Montemor-o-Novo, and with a desire to bring together the cultural and natural resources of the area, MARCA organizes community-based creative activities carried out in two formats: artistic residencies accompanied by masters of knowledge, and workshops. Initial workshops include traditional arts and crafts such as stamping, *Talegos* patchwork sewing (traditional cloth bags for storing bread), basketry, rag dolls, and ceramics. MARCA ADL co-operates with Oficinas do Convento in the ceramics workshops and VAGAR Walking Tours in the promotion of its activities. In addition, MARCA organizes specific activities that merge ecological sustainability issues with arts and crafts.

Município de Beja (municipal government), Beja Criarte (Create Beja). With the development of a local network of traditional creators/makers, Beja Criarte offers a regular stream of workshops of arts and crafts (such as making pottery and chairs), traditional singing (Canto Alentejano), gastronomy, bread-making, and folktales/storytelling. Activities are hosted by the UNESCO Centre for the Safeguarding of Intangible Cultural Heritage in Beja, local restaurants, and a windmill.

Município de Reguengos de Monsaraz (municipal government), Casa do Barro (Clay House). In Reguengos de Monsaraz, the traditional Casa do Barro provides visitors an opportunity to encounter the traditional pottery craft specific to this locale. At this pottery interpretation centre, visitors meet local potters, learn from them about the clay processing cycle and the importance and role of this activity in the local and regional ways of life, and are provided an opportunity to paint or make their own "Alentejo-inspired" plates, with the pieces sent to the participants' homes afterward. While visitors are on-site, they are also introduced to local food preparation traditions relating to this craft and participate in food and wine tastings together with the potters.

Nova Tradição (a culture and tourism grassroots project of the designer Tania Neves), Nova Tradição: Oficinas Têxteis de Base Artesanal (New Tradition: Craft-Based Textile Workshops). Steeped in appreciation for traditional skills and techniques as well as the contemporary "slow fashion" movement, Nova Tradição organizes guided tours and creative workshops related to textile history and the traditional wool cycle, including fabric weaving, plant stamping, and natural dyeing. The project co-operates with the tourism company InEvora, the weavers in Mertola, and local artisans in Évora to offer the wool cycle and wool weaving workshops.

VAGAR Walking Tours (tourism animation agent) has developed Play Évora, a game kit that has been designed for families to support their self-guided, creative discovery of Évora and its many unique historical and artistic features. The kit includes a variety of child- and teenager- (and parent-) friendly activities such as artistic point-of-view games, creative writing, and sketching, along with other incentives to encourage visitors to interact with local residents around themes of heritage and gastronomy. VAGAR Walking Tours also co-operates with the CREATOUR project MARCA ADL.

ALGARVE

Associação Backup (cultural association), AlGharb.Come—do Mar ao Património (AlGharb.Come—from the Sea to Heritage). Based in Vila Real de Santo António, Backup aims to bring local heritage into the future by engaging younger generations in their heritage and sharing it with visitors. The organization is designing and will lead cultural tours that include workshops (held by local residents) on the local fish tinning industry and traditional fishing techniques, the recreation of handicrafts, and the memories and testimonies of the local people.

Associação In Loco (cultural and development association), Tasting Algarve. Located in S. Brás de Alportel, and working with a wide network of independent collaborators, In Loco organizes gastronomic routes throughout the Algarve featuring experiences based on the cultural and natural resources of the region and their strong connection to the Mediterranean diet. For example, they organize cooking and tasting experiences together with local producers who hold the workshops.

Associação Odiana—Associação para o Desenvolvimento do Baixo Guadiana (regional development association), Oficinas do Saber Tradicional (Traditional Knowledge Workshops). The region's aging population has led to the disappearance of secular trades and traditions in the Baixo Guadiana region and it is essential that actions are developed to minimize this decline. In this context, Odiana's artisan-run "Traditional Knowledge Workshops" aim, in a dynamic and interactive way, to revitalize and promote some of the Baixo Guadiana traditions and ancestral know-how. For example, a workshop focusing on traditional salt-making involves participants in all aspects of the process, with visits to active salt-pans and a high-salinity floating pool.

Barroca, Produtos Culturais e Turísticos (tourism and culture company), Tempero (Seasoning). Barroca develops creative tourism activities embedded in the cultural heritage of the Algarve region. These activities promote the application of contemporary visual culture (design and photography) as a vehicle for the promotion and development of local traditions, with a special emphasis on the places of gastronomy. For example, "Sharpening the Perspective," a design experience in small villages, consists of an interpretive tour in the streets of a village inspired by carob, an important local and regional gastronomy element, followed by a creative experience in which participants construct a visual map depicting the processes learned during the activity. Each of Barroca's activities includes tasting experiences with local products.

Centro Ciência Viva do Algarve and **Núcleo do Algarve da Associação de Professores de Matemática** (public agency to promote science to the public in partnership with the Association For Mathematics Teachers), Percusos que Contem (Routes that Count). Within CREATOUR, this partnership is creating an urban walk to transform the city of Faro into a learning space and a place of discovery through challenges that combine science and culture with a playful/game-like approach. Instructions on the route and the challenges will be included in a booklet that will allow participants to explore the city and make use of scientific, artistic, and historical knowledge in a relaxed, informal learning environment that stimulates their creativity.

Eating Algarve Food Tours (tourism company), Loulé Food and Cultural Tour and Tavira Food and Cultural Tour. Immersive tours mixing gastronomy, culture, and heritage are organized in the small towns of Loulé and Tavira guided by a local resident who explains certain objects and traditions related to local eating and culture and creates an immersive experience in the local communities. The pedestrian Food and Cultural Tours are led by local residents who feature places from the daily lives (e.g., restaurants, heritage buildings, and places outside the standard tourism circuit), combining tasting experiences and storytelling.

Espírito da Terra & Co. (a partnership that includes a non-profit association, a parish council, and an organic farm), 3Cs—*Colher, Caminhar, Criar* (Harvest, Walk, Create). Based at an organic garden and farm (that also hosts a Waldorf school) in Boliqueime, participants have the opportunity to visit and explore this territory while participating in three different activities that can be done together or separately: collect organic products from the farm and taste and learn more about them (*colher*); observe the landscape and the natural life surrounding the farm in sensory experience routes (*caminhar*); and create objects from local materials (e.g., wood, leaves, stone, or straw) in creative workshops (*criar*).

Município de Loulé (municipal government), Loulé Criativo: Abertura de Oficinas Tradicionais (Creative Loulé: Opening of Traditional Workshops). The Loulé Criativo initiative encompasses several aspects: creative tourism, including a program of immersion in the local traditional culture; ECOA (Space of Knowledge, Crafts and Arts), with space and equipment for training; and the Loulé Design Lab, which supports the incubation of entrepreneurs related to production and design and creative residences. Within the creative tourism stream, Loulé Criativo has facilitated the development of a variety of traditional heritage-based workshops and activities that intertwine learning about heritage, traditional techniques, and trying them yourself to create an object to take home or a skill to reproduce at home. Within CREATOUR, workshops are held in the areas of coppersmithing, palm weaving, and pottery.

Proactivetur (tourism company), Creative Experiences Program. This tour operator and project manager of the TASA project (Ancestral Techniques Current Solutions) aims to bring strategic innovation to the craft industry by encouraging the use of ancestral craft techniques in modern product design. In partnership with a network of traditional artisans, the organization offers various half-day or full-day workshops on ancestral craft techniques such as cane basket weaving, palm weaving, and making traditional floor tiles.

Tertúlia Algarvia—Centro de Conhecimento em Cultura e Alimentação Tradicional do Algarve (culture and food company), Algarve Cooking Vacations. Within CREATOUR, Tertúlia Algarvia organizes single or multi-day programs for tourists to learn to make various traditional Algarve recipes in hands-on workshops. Each cooking class is preceded by complementary activities such as visits to local producers (such as an organic herb farm, a local greenhouse, an oil press, etc.), markets, historical sites (to learn about related local history and culture), and factories and artisans (where local producers share their know-how).

The Interplay between Culture, Creativity, and Tourism in the Sustainable Development of Smaller Urban Centres

Elisabete Caldeira Neto Tomaz

Introduction

At the end of the 1970s, cities and regions started to mobilize culture in order to boost local economic and social revitalization; since the 1990s, creative-based strategies have also become part of this development agenda. Culture- and creative-led planning is thought to offer distinctive and innovative solutions, products, and activities to attract residents, visitors, and investments. In the new millennium, tourism has become an integral part of these development strategies, with local providers identifying and exploiting distinctive endogenous cultural assets. In the context of the global village and the economic competition its endless possibilities engender, attempts at economic diversification and the expansion of innovation have also brought creativity to the tourism strategy design. Duxbury's chapter in this volume, for example, shines a light on Portugal's approach. Research literature in this field, however, tends to focus on large cities, neglecting the importance and innovative potential of smaller communities. To address that gap, this chapter provides case studies developed in five European small and medium-sized urban centres in intermediate and rural regions. While each of these five examples link culture and creativity with the improvement of tourism, they also seek to contribute to more sustainable forms of community development.

Culture and Creativity in Tourism for Urban Development

Culture and creativity have become linked with urban development policies in the context of the emergence of a "knowledge economy," in which intangible assets determine factors of competitiveness in an increasingly interdependent world (e.g., Garnham 2005; Galloway and Dunlop 2007). The symbolic, socio-cultural, relational, and territorial elements of a place are activated in local development strategies to attract new residents, investors, or tourists, but also to strengthen the sense of belonging, identity, and trust of local communities and stakeholders.

In recent decades, the tourism industry in Europe registered steady growth and a diversification of market segments and destinations. The rise of tourism activity was driven by changes in consumption patterns stimulated by technological advances, the development of low-cost infrastructure and connections, and the evolution of global socio-economic conditions that underpin new motivations and lifestyles. As such, tourism has become a key economic sector and a major contributor to the development and resilience of many cities and regions, given tourism's potential for place-based development and its attractiveness as an alternative source of income and employment (Bellini et al. 2017; Romão and Nijkamp 2017; Weidenfeld 2018). Hence, cities search for greater innovation and specialization in the design of tourism strategies to respond to increasingly qualified and competitive tourism markets and changes in the profile of tourists (Richards and Wilson 2006, 2007; Tan, Kung, and Luh 2013). Their success depends, on the one hand, on a given territory's identification of tangible and intangible assets, and, on the other, on how effectively local resources are brought together to create value, progress, and well-being for the different actors involved in tourism and development (Minguzzi 2006; Minguzzi and Solima 2012).

Today, tourists seek to develop their creative capacity and knowledge through active participation in unique and tailor-made experiences based on a community's culture, identity, and everyday practices (Richards and Wilson 2006, 2007; Tan, Kung, and Luh 2013). Such positive interaction between visitors and residents can strengthen local creative and innovative capacity, promote civic engagement and community participation, and offer new opportunities for learning, socialization, and the expression of locals and visitors alike. Creative industries have thus become an increasingly important component of the urban development agenda, supporting the growth of creative experiences and products in tourist destinations.

Despite the attractiveness of large capitals and metropolises given the regularity, density, and diversity of resources and events, smaller communities can offer a range of alternative cultural and creative experiences and products capable of meeting the evolving interests of tourists and residents. Human-scale, local culture, distinguished built environments, and accessible natural landscapes, as well as the ease of social interaction and sense of identity, are essential factors in the quality and distinctiveness of smaller tourist destinations. Smaller communities are well positioned to provide unique and rewarding experiences for visitors and locals. Examples in this volume range from Portugal to Iceland, from Wales to the Yukon, from the Arctic to South Africa. In this context, many smaller urban and rural communities recognize creativity as a resource capable of generating opportunities for new forms of tourism and developing new modes of governance that can contribute to tourism but also the sustainable development of communities.

This chapter addresses the paucity of research on the creative and innovative potential of small and medium-sized urban centres, examining projects or approaches that attempt to link culture and creativity not only to tourism strategies, but also to a more sustainable approach to urban development and planning.

Creative Tourism Experiences in the Development Processes of Small Communities

The cases summarized here are part of my doctoral research,[1] which focused on small- and medium-sized urban areas[2] located in European "intermediate" and "predominantly rural" regions.[3] This research started with the Creative Clusters in Low Density Areas network (2008–11, European Program URBACT II) and continued under the project COST Action IS1007—Investigating Cultural Sustainability (2011–15).

The EU provides an institutional framework in an attempt to ensure the coherence, effectiveness, and continuity of their policies and actions,[4] although it does not have explicit legal competencies in spatial planning or the cultural field. These are the responsibility of the national states, regions, and/or municipalities. However, this does not mean that the EU has no influence in these areas; this influence is felt, for instance, through sectoral legislative and funding programs that determine much of the political agenda and strategies adopted by member-state governments, as well as by local authorities and cultural organizations that depend on these programs.

The five selected urban centres described in this chapter have distinctive circumstances (socio-economic, political, historical, cultural) that are linked to their national contexts[5] as well as to the cultural policy regimes[6] to which they belong.

Český Krumlov (Czech Republic)

The Český Krumlov District, with a total of 61,226 inhabitants (CZSO 2018), is one of the smallest districts of the South Bohemia region. Classified as predominantly rural according to Eurostat (2012), it is composed of low-density urban settlements, and it benefits from a natural and diverse ecosystem.

The district's eponymous capital has only 12,981 inhabitants (CZSO 2018). It is situated 25 kilometres south of the regional capital, České Budějovice, 220 kilometres south of Prague, and 70 kilometres north of Linz, in Austria. This geographical location shaped the development of the city, which has flourished as an important craft trade and cultural centre since medieval times. The surrounding landscape and urban structure of Renaissance and baroque bourgeois houses, framed by the Vltava River and surmounted by the castle complex, has become the engine of the local development strategy and an important cultural destination.

In 1963, the town was declared a Municipal Reserve, and in 1989 the castle was designated a National Monument. After a period of decline, the urban renewal process began in 1971; this accelerated in the 1990s after the fall of the communist regime, when the Český Krumlov government started acquiring large pieces of real estate in order to restore the historic centre. Administered by the Český Krumlov Development Fund, this process acquire a new dimension with the inclusion of the Historic Centre of Český Krumlov in the UNESCO World Heritage List in 1992. In 1995 and 1996, 80 per cent of the old townhouses were rebuilt and various regulations on tourist infrastructure, traffic, conservation measures, and use of public spaces were generated to aid the promotion and growth of tourism, which has become a vital source of income for the municipal government.

As part of this strategy, local authorities, in collaboration with educational, cultural, and civic institutions, have promoted a robust agenda of cultural events that have attracted large numbers of attendees; these include the Five-Petalled Rose Celebrations (a Renaissance festival); the Festival of Baroque Arts; and sports competitions such as Rally Český Krumlov. Moreover, these cultural activities were complimented by various theatres (such as the Town Theatre and the Revolving Auditorium) whose programming perpetuate a long and rich dramatic tradition in the region. Museums and galleries (e.g., the Regional Museum) have also drawn many tourists, although the biggest visitor attraction is the town's varied architecture.

The Český Krumlov State Castle, a major tourist attraction, is an example of how local historical conservation has led to the investigation and preservation of regional authenticity through the use of historical artisanal building practices and traditional materials. Its administration is a key player in the local development strategy,[7] promoting creative experiences for visitors that range from gastronomic delights to musical and theatrical performances. A good example is the Baroque Nights series, which tries to recreate in an authentic way the atmosphere of noble Czech families—in this case, the Eggenbergs and the Schwarzenbergs—while also encouraging the development of local specialized skills and related businesses.

Another restored cultural venue is the photography studio and home of Josef and František Seidel, which was refurbished with tourism revenues and EU funds and now stands as a unique testimony to the region's late-nineteenth-century social and cultural history.[8] It is not just the well-preserved collection depicting the surrounding natural landscape and the celebrations and routines of the local community that make this place so unique for those who visit it. The physical and emotional feeling that transmits endures through the photographs taken, even today, of many generations of families who return to the studio.

Český Krumlov is a paradigmatic example of a UNESCO city, with its well-preserved heritage surrounded by a picturesque natural landscape that defines the town's distinctive character: the *genius loci*, the same elements that have attracted many aristocratic families, artists (such as painters Wilhelm Fischer and Egon Schiele), and visitors over the centuries, and which locals seek to maintain today. Its exceptional reputation has increased the number of annual visitors (to more than 1.5 million in 2018) and transformed the city into an important site of tourist consumption (Ashworth and Page 2011). For this reason, it is very attractive to investors, while locals have seen the value of local properties and goods rise as well. This situation has pushed many permanent residents to the suburbs, thus leaving the historical centre dominated by tourist accommodations and business, some of which have little to do with local traditions and products.

On the other hand, despite these criticisms, the restoration of local heritage, combined with the influx of tourism, has fuelled the revitalization of the historic city centre and local cultural traditions and events, reinforcing the community's pride and creative participation. Other challenges faced by locals derive from the management of conservation commitments; these guarantee the maintenance of the UNESCO seal but also imposes difficulties in adapting to the needs of contemporary life.

Despite the growth in tourism in Český Krumlov, the sector faces some challenges related to the need to diversify its offerings and extend visitors' stay throughout the year. Some pertinent questions are also raised about authenticity and the concept of heritage as a living testimony of a community, and, consequently, about the sustainability of a culture-based development strategy that is subordinated to tourism. Our next example, York, while it also attracts thousands of visitors to its events, is less dependent on tourism because of its size and relative diversity of resources.

York (United Kingdom)

The City of York, a unitary authority (single-tier local government) of 208,163 inhabitants (City of York Council 2019), is located in the county of North Yorkshire, England, in the vicinity of the large urban conurbation of Leeds and less than two hours from London. Classified as an intermediate region (Eurostat 2012), the densely populated city centre is delimited by an external ring road that separates it from a wide rural landscape dotted with small settlements.

With more than two thousand years of urban history, York gained importance firstly as a centre of trade and textile manufacturing and later as a centre for railway and confectionery industries. By the end of the twentieth century, local leaders, confronted with the decline of their traditional industries but also challenged by the changes in national policy guidelines related to heritage preservation and urban planning, adopted a development strategy focused on the study and conservation of the city's unique historical legacy.

York is one of five historical centres in England designated as an Area of Archaeological Importance, and since that designation, the York City Council has acted to preserve the city's historic character by restoring listed buildings, designating pedestrian areas, and excavating and protecting archaeological sites, as well as creating various cultural venues and collections supported by an extensive program of events and collaborations with the private sector. Further, the growth of the global tourism industry has gradually encouraged a "postmodern approach" to heritage in York, transforming it into a cultural commodity for individual and tourist experience (e.g., Hewison 1987; Urry 1995; Meethan 1996; Selby 2004). As one local stakeholder explained, if until the 1990s, "tourism, conservation and archaeology were perceived as problems that [the municipal government] had to manage. . . . After that, there was an increasing awareness that the basis of York's economy is its heritage" (Tomaz 2018). In fact, the visitor economy has become a strategic priority in the city's development. In 2018, 8.4 million people visited York, which resulted in £765 million worth of spending and supported more than 24,000 jobs in the tourism industry.

One of the leading organizations in this local development strategy is the York Archaeological Trust,[9] which has been responsible for archaeological excavations in the areas subject to intervention since the 1970s within the framework of local regeneration plans. The organization provides fieldwork experience for professionals, researchers, and volunteers from around the world. Based on Coppergate's archaeological excavations, begun in 1972, the York Archaeological Trust created the Jorvik Viking Centre, which presents practical and sensory experiences of Viking life as it existed in York (then called Jorvik) between 866 and 954. Subsequently, the Trust inaugurated the DIG: An Archaeological Adventure, in which children (both locals and tourists) can enjoy hands-on archaeological experiences; these are complimented by visits to the Barley Hall house and by the Richard III and Henry VII Experiences, which aim to divulge the history of medieval York. If these efforts have contributed to a better understanding and recognition of the need to preserve York's historical legacy, the revenues from related tourism activities have enabled the Trust to develop research, education, and non-profit activities that have in turn brought benefits to local businesses and organizations.

Local authorities has also invested in the creation of a recognized education, science, and technology cluster based on prestige higher education institutions and the attraction of financial and commercial services. More recently, they launched a successful bid to become a UNESCO Creative Cities City of Media Arts, which was approved in 2014. The aim was to strengthen the connection between the creative and cultural sectors and sustainable urban planning and development, including by providing experiences for longer-term and higher-spending visitors. This designation reflects the focus on culture that local partnerships composed of representatives from the public, business, and voluntary sectors have long brought to debates around local strategic development.

In 2015, the York City Council created Make It York, a company responsible for the development of culture, business, and tourism. This strong and interconnected urban development strategy stands out in comparison to other places. But, like Jyväskylä, partnerships between different sectors, along with the role played by educational institutions, are determining factors in the development of local strategies.

Jyväskylä (Finland)

The city of Jyväskylä, located 270 kilometres northeast of the Finnish capital of Helsinki, is the capital of the Central Finland region. The surrounding municipality, which Eurostat defines as predominantly rural, comprises a vast hinterland of lakes, forests, and hills, but more than half the region's population live in the city. In fact, although the municipality exceeds 141,305 inhabitants (Statistics Finland 2019), a large part of its population lives within walking distance of the city centre, which helps maintain the atmosphere of a small town. It has a very young population, of which 45,000 are schoolchildren and students. This is thanks principally to the existence of high-quality and diverse health and education providers, which are especially attractive to young families.

Referred to as the "Athens of Finland" because of its internationally renowned educational cluster, the Jyväskylä region has bet since the 2000s on a new development strategy, encapsulated in the name "Human Technology City," that is based on a knowledge-intensive economy and investment in innovation and technology in order to create opportunities in new sectors such as electronics, energy, environment, well-being, and nanotechnology. However, the 2008 global financial crisis and the closure of the local Nokia Research Center in 2009 brought new challenges for the city. Its educational and research system nonetheless continue to support an innovation ecosystem model "in which resources in companies, among citizens and in the public sphere are put to good use to create genuine synergies" (Hautamäki and Oksanen 2015, 96). This model shapes local strategies, which in turn, following national and European trends, and fosters creativity and culture as drivers of local development in connection with the values of well-being and wellness. Both concepts are represented by the Finnish word *hyvinvointi* (which refers to both the notion of well-being and wellness), and they reflect a conception of local identity and culture that local authorities and companies in Jyväskylä have sought to promote with the development of wellness tourism.

Under the brand Sauna from Finland, an extensive networks of stakeholders from different sectors (from sauna manufacturers and service providers, to health-care providers and technology developers) and supported by the local development agency are developing new services, products, and events aimed at amplifying this typically Finnish tradition.[10] The interaction between the high-tech creative industries and tourism providers was underlined by the Regional Council as a way of creating opportunities to add value to traditional activities and to support the lifestyles and values of local communities, especially in the countryside.

Jyväskylä is already recognized as a destination for fairs and congresses, as well as sporting and cultural events like the popular Neste Oil Rally, the Jyväskylä Arts Festival, and the Triennial Graphica Creativa, which are supported by a range of cultural amenities. Furthermore, the work of the internationally famous architect Alvar Aalto (1898–1976), who considered Jyväskylä his home town, is another distinctive factor in, and a major contribution to, the quality of the city's urban fabric, attracting many researchers and tourists curious about his work.

Local development therefore focuses on innovation and creativity as they relate to local cultural values. These are seen as engines of economic development, but also as qualities that improve the delivery of health and social welfare services. This approach is also seen as a way to enhance the region's attractiveness and to differentiate it from other tourist destinations. Intersectoral collaboration, as in the case of York, is essential to the sustainability and scale of this strategy. And as we will see, health and wellness are also relevant to the development and diversification of tourism in Óbidos.

Óbidos (Portugal)

The Municipality of Óbidos, on the west coast of Portugal, is a predominantly rural region (Eurostat 2012). It is 141.6 square kilometres and is dotted by several small villages in a vast natural landscape that includes an ecosystem of lagoons that ends in the Atlantic Ocean: a popular spot for bird watching, as well as for sports and recreation.

The small town of Óbidos is the seat of this municipality of only 11.850 inhabitants (Statistics Portugal 2019), and is located about eighty kilometres north of Lisbon, the Portuguese capital, on the route to the pilgrimage centre of Fátima. Recognized internationally for its iconic medieval walled village, with its narrow streets and whitewashed houses, it was a refuge for nobles and artists, who have left a unique symbolic and architectural legacy.

After a period of abandonment and deterioration at the beginning of the twentieth century, the development of the tourism industry began with the classification of the Óbidos Castle as a National Monument in 1910 (subsequently extended to the entire walled city centre in 1951) and its designation as a Tourist Resort in 1928. An extensive restoration program was carried out by the General Directorate of National Buildings and Monuments, working with the Municipality in the reconstruction and regulation of the built environment to impose a harmonized image of the city. In addition, new cultural facilities have been created, such as the Contemporary Art Gallery and the Municipal Museum (both founded in 1970), and various local traditions and festivities have been recovered and reinvented, creating a crystallized narrative of the city's medieval past (Soares and Neto 2013).

In 2002, the Municipality initiated a new development strategy and rolled out a powerful marketing campaign—the "Creative Óbidos" brand, which links culture, tourism, and the economy in the context of local regeneration and economic diversification plans. Óbidos was soon recognized for its annual agenda of thematic events supported by a network of cultural facilities that attract thousands of tourists to the small town. In the first decade of the 2000s, the perception of Óbidos as a museum town changed drastically as locals developed new creative skills, innovative businesses, and cultural activities in an effort to (re)interpret local tangible and intangible assets. A good example is the International Chocolate Festival of Óbidos. During this event the village is invaded by tourists, who, in addition to the revenues gathered by the Municipality from the festival's ticket sales and sponsorships, also bring economic gains for local companies during the low season. The festival welcomes local primary students, whose poems are engraved on chocolate bars, as well as students and teachers from the West School of Hospitality and Tourism, who show their skills in the construction of chocolate sculptures. The festival has also stimulated the emergence of creative tourism activities and businesses based on related endogenous products, such as the Oppidum Company's Ginja de Óbidos brand. As a result of the success of the gourmet products presented at the festival combining locally produced sour cherry liqueur with chocolate, this local company has expanded its commercial offer. Indeed, it now provides visits and tasting experiences at its manufacturing site (Pimpão 2016).

Since 2009, the Municipality has activated an action plan focused on attracting creative talent and businesses, supported by a range of infrastructures (creative residences, collaborative spaces, incubators, etc.), and on promoting diversified tourism based on the high quality of life enjoyed in the area as a result of its excellent cultural and natural environment. The development strategy was also strengthened with the participation of the Municipality in national and European networks that extend the visibility and recognition of Óbidos, thereby giving it a larger scale and critical mass.

As we have seen with York and Jyväskylä, sustainable development is most effective when public and private entities work together. Óbidos' designation as a UNESCO Creative City of Literature is one such an example, in which the local development strategy combines tourism-derived economic benefits with socio-cultural development. With the support of the Municipality, the company Ler Devagar[11] (which manages editorial funds) launched the Óbidos Literary Town project in 2015. This included the successful organization of the International Literary Festival FOLIO and the opening of several bookstores in unusual spaces (markets, galleries, churches), thereby reaching new audiences, partnerships, and increasing community engagement. This process culminated with the designation of Óbidos as a UNESCO Creative City of Literature in December 2015.

Óbidos, like Český Krumlov, faces great difficulties in maintaining a balance in its walled village between the lives of residents and demands made by the many tourists who into the region, thereby connecting a rural territory to the urban narratives of the cultural and creative economy. Despite this, the local development strategy has helped to reshape Óbidos by empowering local stakeholders and attracting new actors and projects that keep the community alive. One of the least visible foundations of the local strategy was the implementation of a municipal education model based on creativity and innovation that transcends the classroom. As we will see in the following section, this is also the case in Reggio Emilia.

Reggio Emilia (Italy)

Reggio Emilia (also known as Reggio nell'Emilia) is, with 172,124 inhabitants (UrbiStat 2019), one of eight provincial capitals of the Italian region of Emilia-Romagna. Integrating an intermediate region, according to the Eurostat (2012), it combines a rich rural landscape with densely populated villages, and it offers various places of natural, historical, and socio-cultural interest for travellers looking to experience an authentic Italian lifestyle.

The local military castrum, built by the Romans near the Via Emilia at the beginning of the second century BCE, soon became a thriving city that in the centuries since has been a leader in the political, socio-economic, and cultural life of the region. Historically, the local economy has been primarily agricultural (from large industrial units to small and medium-sized companies); however, since the 1900s, industrial activity has increased, initially in the mechanical sector for agricultural equipment, as well as construction and manufacturing; more recently, the region has seen growth in the highly specialized mechatronics area.

The cooperative movement with a long tradition in the city, plays a very significant role not only from an economic standpoint but also from a socio-cultural perspective as well as in terms of building an atmosphere of solidarity and the protection of civic values. This effort was supported by several associations, notably in the preschool system and as part of the pedagogical method put into practice after the Second World War, which has been supported by the efforts of teachers and the community at large. The so-called Reggio Approach is today an internationally recognized model for early childhood education, one sustained by the Loris Malaguzzi International Centre and run by the Foundation Reggio Children. The project coordinates annual educational events and conducts research with institutions and educators from more than 120 countries around the world. In addition, it holds workshops and ateliers to engage children and adults in creative and exploratory activities, and it produces educational products with private companies.

The Foundation Reggio Children has developed a creative and innovative experience that meets the vision of local authorities in promoting the development of a knowledge economy, with infrastructural support provided by the Area Nord regeneration plan, such as Tecnopolo (managed by the University of Modena and Reggio Emilia). This abandoned industrial area was transformed into a research and knowledge-transfer centre, the Parco dell' Innovazine (Innovation Park), which seeks to explore the potential and synergies that exist in the region.

The region already offers a variety of attractions and experiences related to local history, art, gastronomy, and natural scenery. Nevertheless, Reggio Emilia is not especially known as a tourist destination; instead, it is essentially viewed as either a business destination or as a research hub for those studying the Reggio Emilia educational experience. The Municipality has promoted tourism as an economic sector capable of offering new opportunities to the region. But the question remains: In a country rich in artistic and cultural heritage and with many competing destinations for tourists, how can Reggio Emilia leverage its tangible and intangible heritage?

As a way of restoring the city's historic centre and promote cultural tourism, the Municipality renovated the historic buildings in the urban centre and established a network of cultural venues, such as Officina delle Arti, an interdisciplinary space that hosts an exchange program in contemporary art; or the Palazzo dei Musei, a civic historical museum. The organization of events and festivals such as European Photography, which every year attracts thousands of visitors, is also part of the local strategy. And yet, while these initiatives are promising, they are identical to so many other cases, as we have seen before.

Having a more diversified economic base, as is the case in Jyväskylä and York, Reggio Emilia is less dependent on tourism revenues. Thus, it can benefit from a niche market approach, one based on genuine local products for more sustainable tourism. This will help the region to align its spatial planning efforts with the local culture of innovation and co-operation, so that it does not fall into the resident/visitor division that the historic centres of Český Krumlov and Óbidos have experienced.

Final Comments

As Kathleen Scherf mentions in this volume's introduction, smaller communities and their local governments, facing the decline of their traditional economic bases, must adopt proactive forms of intervention and new development strategies. Although they may choose different models and narratives, they must involve non-governmental actors and new modes of financing, given the reductions in state support and the growing pressure to justify the provision of public services.

In this context, the development of tourist activity can be a useful lever. As Nancy Duxbury points out in chapter 1, tourism can offer advantageous opportunities for these smaller communities if they are willing to focus on creating niche markets that leverage their tangible and intangible endogenous resources and to realize that sustainability will require small-scale sustainable tourism practices. Given the characteristics of smaller cities, especially in intermediate and rural regions, and in line with Greg Richards's observations in the conclusion to this volume, it is necessary to overcome their relative disadvantage by adopting different approaches than those employed by metropolises. The objective is to create an environment conducive to culture-based creativity, using existing and local resources efficiently and innovatively, including the location itself, and the close proximity between local actors. However, as the examples in this chapter show, also essential to sustainable socio-cultural and economic development is the promotion of synergies and networks to achieve greater scale, visibility, and critical mass. Deploying culture and creativity is fundamental to the creation of rewarding and innovative experiences and products for visitors, promoters, and communities. The location, history, and characteristics of a place and its communities influence and define which activities are successful. Tourism experiences in the five communities studied in this chapter all benefit from the proximity of rural and natural settings, and in at least three of them, the connection between visitors and host communities, which come together via experiences based on endogenous assets, stimulates creative expression and learning.

These experiences based on creativity and culture have also promoted the development of new sectors of activity, along with partnerships that enhance the resilience of more traditional sectors, with multiplier effects in several areas. They also promote the engagement and cohesion of communities and their values, practices, and knowledge, as well as the development of new modes of governance that answer not only to tourist demand but also address sustainability objectives, contributing positively to socio-economic progress and the protection and preservation of natural, social, and cultural resources.

The intensification of tourism activities presents challenges, especially when an area's economic base is less diversified or its tourism strategy is an "add-on" rather than a genuine engagement with the cultural plan of the city and its everyday community practices. The cases studied in this chapter highlight the differences in how the various municipalities used culture and creativity in their own urban planning, combining them with the possibilities presented by the tourism industry. Accordingly, they enable us to evaluate how effective these strategies were when it came to reinventing these cities and seeking new forms of economic revitalization, community building, and sustainability in a "post-industrial" or "post-rural" society.

NOTES

1 Supported by the Portuguese Foundation for Science and Technology.

2 This designation intends to describe municipalities with 5,000 to 250,000 inhabitants, which also corresponds to a specific set of functional relations within the territorial system and between spatial scales with consequences for the formulation and development of political strategies.

3 This classification is made by Eurostat according the new urban-rural typology for NUTS 3 Regions, available at https://ec.europa.eu/eurostat/statistics-explained/index.php?title=Territorial_typologies_manual_-_urban-rural_typology#Links_to_other_spatial_concepts.2Ftypologies.

4 See article 13 of the Consolidated Version of the Treaty on European Union, available at https://eur-lex.europa.eu/resource.html?uri=cellar:2bf140bf-a3f8-4ab2-b506-fd71826e6da6.0023.02/DOC_1&format=PDF.

5 We consider different types of EU countries according to their state systems, competencies between different levels of government, and local authorities' autonomy based on a study carried out by Ismeri Europa and Applica (2010).

6 The cultural policy regimes reflect the historical, political, and institutional configurations of each country and describes the "socio-economic structure, their internal hierarchies and the conceptions of art and culture that prevail within them" (Dubois 2013, 1).

7 The project to restore the southern facade of the Upper Castle in the Český Krumlov State Castle and Chateau received the 2008 European Union Prize for Cultural Heritage/Europa Nostra Award.

8 In 2009, the Fotoatelier Seidel received the Gloria Musealis, the highest Czech award for a museum.

9 York Archaeological Trust is an independent charity; its website is available at http://www.yorkarchaeology.co.uk/.

10 The network's multilingual website is available at https://saunafromfinland.com.

11 Ler Devagar (literally "slow reading") is a public limited liability company.

References

Ashworth, Gregory, and Stephen J. Page. 2011. "Urban Tourism Research: Recent Progress and Current Paradoxes." *Tourism Management* 32, no. 1: 1–15.

Bellini, Nicola, Francesco Grillo, Giulia Lazzeri, and Cecilia Pasquinelli. 2017. "Tourism and Regional Economic Resilience from a Policy Perspective: Lessons from Smart Specialization Strategies in Europe." European Planning Studies 25, no. 1: 140–53.

City of York Council. 2019. "York Profile". Business Intelligence Hub. Accessed 28 September 2020. https://data.yorkopendata.org/dataset/2a844263-5c89-4fe1-95e9-ee5ae289598d/resource/ec0adf86-63d8-492e-b117-fa17a68c0610/download/york-profile.pdf.

CZSO. 2018. "Population of Municipalities—1 January 2020—Jihočeský Kraj/Region." Accessed 8 July 2020. https://www.czso.cz/csu/czso/population-of-municipalities-1-january-2019.

Dubois, Vincent. 2015. "Cultural Policy Regimes in Western Europe." In *International Encyclopedia of the Social & Behavioral Sciences*, 2nd ed., edited by J. D. Wright, 460–465. London: Elsevier.

Eurostat. 2012. *Eurostat Regional Yearbook 2012*. Luxembourg: Publications Office of the European Union.

Galloway, Susan, and Stewart Dunlop. 2007. "A Critique of Definitions of the Cultural and Creative Industries in Public Policy." *International Journal of Cultural Policy* 13, no. 1: 17–31.

Garnham, Nicholas. 2005. "From Cultural to Creative Industries: An Analysis of the Implications of the 'Creative Industries' Approach to Arts and Media Policy Making in the United Kingdom." *International Journal of Cultural Policy* 11, no. 1: 15–29.

Hautamäki, Antti, and Kaisa Oksanen. 2015. "Sustainable Innovation: Competitive Advantage for Knowledge Hubs." In *Orchestrating Regional Innovation Ecosystems: Espoo Innovation Garden*, edited by Pia Lappalainen, Markku Markkula, and Hank Kune, 87–102. Espoo, FI: Aalto University in co-operation with Laurea University of Applied Sciences and Built Environment Innovations.

Hewison, Robert. 1987. *The Heritage Industry: Britain in a Climate of Decline*. London: Methuen.

Hjalager, Anne-Mette, Henna Konu, Edward Huijbens, Peter Björk, Arvid Flagestad, Sara Nordin, and Anja Tuohino. 2011. *Innovating and Re-branding Nordic Wellbeing Tourism. Final Report from a Joint NICe Research Project*. Oslo: Norden: Nordic Innovation Centre.

Ismeri Europa and Applica. 2010. *Distribution of Competences in Relation to Regional Development Policies in the Member States of the European Union: Final Report*. Brussels: DG Regio, European Commission. http://ec.europa.eu/regional_policy/sources/docgener/studies/pdf/2010_distribution_competence.pdf.

Meethan, Kevin. 1996. "Consuming (in) the Civilized City." *Annals of Tourism Research* 23, no. 2: 322–40.

Minguzzi, Antonio. 2006. "Network Activity as Critical Factor in Development of Regional Tourism Organization. An Italian Case Study." In *Tourism Local Systems and Networking*, edited by Luciana Lazzaretti and Clara S. Petrillo, 257–85. Oxford: Elsevier.

Minguzzi, Antonio, and Ludovico Solima. 2012. "Relazioni Virtuose Tra Patrimonio Culturale, Turismo e Industrie Creative a Supporto Dei Processi Di Sviluppo Territoriale." Proceedings of the conference *Atti del XXIV Convegno Annuale Di Sinergie*: 641–53.

Pimpão, Marta Beatriz Oliveira. 2016. "Consolidação de Processos Chave Do Sistema de Gestão Da Qualidade (Da Empresa Dário Albano Zina Pimpão Lda–Ginja de Óbidos Oppidum)." MA thesis, Escola Superior de Turismo e Tecnologia do Mar e o Instituto Politécnico de Leiria.

Richards, Greg, and Julie Wilson. 2006. "Developing Creativity in Tourist Experiences: A Solution to the Serial Reproduction of Culture?" *Tourism Management* 27, no. 6: 1209–23.

———, eds. 2007. *Tourism, Creativity and Development. Contemporary Geographies of Leisure, Tourism and Mobility.* London: Routledge.

Romão, João, and Peter Nijkamp. 2017. "Impacts of Innovation, Productivity and Specialization on Tourism Competitiveness: A Spatial Econometric Analysis on European Regions." *Current Issues in Tourism* 22, no. 10: 1150–69.

Selby, Martin. 2004. *Understanding Urban Tourism: Image, Culture and Experience.* London and New York: I. B. Tauris.

Soares, Clara Moura, and Maria João Neto, eds. 2013. Óbidos *da "Vila Museu" a "Vila Cultural": Estudos de Gestão Integrada de Património Artístico.* Casal de Cambra, PT: Caleidoscópio.

Statistics Finland. 2019. "Key Figures on Population by Region: Jyväskylä, 2018." Accessed 8 July 2020. http://www.stat.fi/tup/suoluk/suoluk_vaesto_en.html#Population%20 data%20by%20region.

Statistics Portugal. 2019. "Annual Provisional Resident Population Estimates." Accessed 8 July 2020. https://www.ine.pt/xportal/xmain?xpid=INE&xpgid=ine_ indicadores&indOcorrCod=0008273&xlang=pt.

Tan, Siow-Kian, Shiann-Far Kung, and Ding-Bang Luh. 2013. "A Model of 'Creative Experience' in Creative Tourism." *Annals of Tourism Research* 41 (April): 153–74.

Tomaz, Elisabete. 2018. "Culture and Urban Development Policies. Beyond Large Metropolis." PhD diss., School of Sociology and Public Policies, ISCTE-IUL.

UrbiStat. 2019. "Municipality of REGGIO NELL'EMILIA: Demographic Statistics." Accessed 18 September 2010. https://ugeo.urbistat.com/AdminStat/en/it/ demografia/dati-sintesi/reggio-nell-emilia/35033/4 .

Urry, John. 1995. *Consuming Places.* International Library of Sociology. London: Routledge.

Weidenfeld, Adi. 2018. "Tourism Diversification and Its Implications for Smart Specialisation." *Sustainability* 10, no. 2: 319.

The Role of Cultural Festivals in Regional Economic Development: A Case Study of Mahika Mahikeng

James Drummond, Jen Snowball, Geoff Antrobus, Fiona Drummond

Introduction

The Mahika Mahikeng Music and Cultural Festival takes place over three days in Mahikeng, the capital of South Africa's North West province (named Bokone Bophirima in the local Setswana language), in early December. Although it is held around the same time as an older event, it began in its current form in 2015 as a government-driven local economic development scheme. And while previous festivals were organized by local promoters and producers and featured local, national, and international artists, the Mahika Mahikeng Festival departed from this local community-led initiative. It is described by the organizers as a "Music and Cultural Festival which provides a platform for product position and paradigm innovation for the creative industries sector in Bokone Bophirima." (Mahika Mahikeng n.d.). An unusual feature for a medium-sized town, Mahikeng has a wide variety of large performance spaces that are already equipped with seating, lighting, and sometimes sound equipment. The availability of these venues provides an opportunity for expanding the festival since, unlike in other medium-sized towns, Mahikeng is not constrained by supply-side limitations. This is uncommon in smaller communities, which are usually disadvantaged by a lack of density and dispersed co-location of players in the cultural and creative sectors. In the case of Mahikeng, a creative cluster already exists; Scherf points to the importance of such clusters in the introduction to this volume.

Cultural events like the Mahika Mahikeng Festival can contribute to regional development in a number of ways, including economic (attracting tourism, marketing the area and its environs, stimulating local businesses), social, and intrinsic (cultural) impacts. However, the 2016 edition of the Mahika Mahikeng Festival faced some challenges in terms of organization and audience attraction. Using the *Framework for the Monitoring and Evaluation of Publically Funding Arts, Culture and Heritage* (henceforth *M&E Framework*) created by the South African Cultural Observatory (SACO 2016), this chapter analyzes the 2016 Mahika Mahikeng Festival in terms of its current impacts and future potential for contributing to the development of the creative sector and the attraction of creative tourists. Results showed that, while the festival was largely successful in attracting a local audience who recognized the potential value of the event, organizational factors, such as marketing, the availability of transport, and the timing of the event, have constrained its ability to attract an audience of creative tourists. Most notably, the organizers made little effort to involve local stakeholders, thus missing the opportunity to úse the festival as a means to develop a cultural place identity to entice creative tourists who desire an authentic engagement with endogenous creative activities. The festival thus failed to meet some of the necessary conditions—outlined by Scherf in her introduction to this volume—for a host community to encourage creative tourism. In particular, in Mahikeng there was a distinct lack of collaboration between the local community and the government organizers during the planning phase. This threatens the sustainability of the festival.

Literature Review: Festivals, Place, and Local Economic Development

Traditionally, festivals were associated with the celebration of some aspect of the community in which they took place, such as cultural customs, history and heroes, agriculture (harvest festivals), or religion. However, in recent times, they have also increasingly become vehicles for local economic development (LED) and place marketing (Visser 2005, 155; Van Zyl

2011, 182). Ma and Lew (2012) attribute the great increase in the number of cultural festivals in China to both economic and social transformations and to a desire to "build place images and boost regional economic development" (14).

Very similar patterns can be identified in South Africa where, post-apartheid, there has been a phenomenal increase in the number of festivals, justified on artistic, social, and cultural grounds as well as economic ones (Visser 2005, 155; Van Zyl 2011, 181). In an early article on the rise of South African festivals, Visser (2005, 166) notes that, in terms of spatial distribution, festivals occur very unevenly, with the majority taking place in wealthier, more populous provinces (Western Cape, Gauteng, and KwaZulu-Natal) with larger urban centres. He also notes that they tend to occur in two main time periods, both of which coincide with school holidays: the first in March/April, and the second in September/October. Moreover, newer festivals were specifically timed not to coincide with more established ones, which is evidence of the organizers' perceptions of increasing competition (167).

Cudny (2014, 132) attributes the increase in the number of festivals to the rise of experience tourism. For Ma and Lew (2012), tourists are "semiotic armies who move around the world to gather signs of local identity . . . participating in cultural sampling through which they collect signs that typically represent the cultures of destinations" (18). Such tourists, through their spending, provide LED opportunities, not only through the short-term economic impacts of their spending, but also as a way of developing the place identity of the host city, as well as boosting job creation through longer-term development of the cultural industries and cultural infrastructure (Van Aalst and Van Melik 2012, 196). Duxbury and Campbell (2011, 114) argue that festivals, like other areas of the cultural and creative industries, can be effective LED tools in rural communities as well as in cities. In fact, the unique local histories and traditions of rural areas can be an advantage in marketing cultural festivals. Creative tourism, in particular, emphasizes visitor immersion in local cultural and creative industries, with a view toward regenerating sustainable cultural development. Chapters in this volume by Duxbury, Tomaz, Aquino and Burns, and Richards reflect the importance of local collaboration in developing events that successfully attract creative tourists.

Marketing is a challenge faced by festival organizers in contexts where competition for both audiences and artists is increasingly fierce. Van Zyl (2011, 187) argued that, for the long-term success of festivals, it is important that each event is positioned in a unique way so that a niche market of loyal supporters is developed. To do this, careful research into all the factors that contribute to the audience's experience must be undertaken. The motivation for attending a cultural festival is necessarily related to the quality, price, and variety of cultural offerings, but other considerations, such as a desire to escape from one's everyday life or to socialize with family and friends, as well as activities such as food and shopping, and facilities such as accommodation and transport, can also be very important in determining the ability of an event to attract tourists (Ma and Lew 2012, 13; Van Zyl 2011, 182).

Ma and Lew (2012, 14) caution that too much of a focus on the tourist aspects of a cultural festival can result in commodification and the erosion of the event's authenticity. They argue that "festivals grounded in a long history of place-specific traditions are more likely to instill tourist expectations that include experiences of authenticity and locally distinct cultures" (14). There is no doubt that, particularly for those festivals that are strongly linked to the history and customs of a certain place, geography matters—both in maintaining the "authentic" experience linked to intrinsic cultural values, as well as in developing a distinct festival brand, or niche, that differentiates an event from its competition and attracts tourists.

For some festivals, the history and culture of the place in which they occur are intimately linked to the development of their unique cultural offerings. These are what Ma and Lew (2012, 16) refer to as "Local Heritage Festivals," which are strongly grounded in local identity and traditions. While some may include contemporary cultural offerings, these festivals are for the most part highly "place-specific." In contrast, some festivals are "place-nonspecific" in that their cultural offerings are linked to national or global trends rather than to the specific location in which they occur. The point that Ma and Lew make is that the extent to which a festival is "rooted in place" needs to be carefully considered when designing and marketing the event (16). For example, if the festival's marketing emphasizes the identity and traditional heritage of a place, this needs to be reflected in the

Table 3.1. Festival Types and Links to Place Specificity

TYPES OF FESTIVALS	DEGREE OF PLACE SPECIFICITY	CHARACTERISTICS
1. Local heritage festivals	Highly place-specific & linked to local identity	Mostly local market focused
		Vernacular and non-tourism in origin
		Celebrating local historical events, figures, and cultural traditions
2. Local modern festivals	Less place-specific, but still strongly linked to local identity	Mostly local market focused
		Tourism and image development in origin
		Celebrating local food products, artists, or other characteristics
3. National heritage festivals	Place-nonspecific	Local to national market focused
		Vernacular/historical and non-tourism in origin
		Celebrating national events, figures, and shared traditions
4. Global modern festivals	Place-nonspecific	Local to international market focused
		Tourism and image development in origin
		Celebrating international arts and business activities

Source: Adapted from Ma and Lew (2012, 21).

cultural offerings—otherwise the event risks being perceived as lacking in authenticity and uniqueness and is unlikely to be sustainable in the long-run. Place-specific local heritage and local modern festivals also have different target markets than place-nonspecific national heritage and global modern festivals (see table 3.1).

For place-nonspecific festivals, Van Aalst and Van Melik (2012) demonstrate that place really does not matter. They studied the North Sea Jazz Festival, which was moved from The Hague, where it started, to Rotterdam, some twenty miles away, in 2006. They acknowledge that, for host cities, festivals can be valuable on a number of fronts: as a showcase for the city; as a "creative destination and breeding ground for talent" (197) (that is, as a contributor to the longer-term development of the cultural and creative industries); and through the economic impact of tourist and organizer spending. However, they also note that, with modern festivals that celebrate international, place-nonspecific culture, the need for a festival to happen in a specific location is reduced, making them "more footloose" (198).

Certainly, in the case of the North Sea Jazz Festival, Van Aalst and Van Melik's findings show that the festival's move to Rotterdam had a very limited impact on tourist numbers or their overall profile, other than a drop in the proportion of visitors from The Hague and an increase in visitors from Rotterdam (2012, 202). Since most attendees were attracted by the music rather than the local culture, they conclude that place was only important in fulfilling certain characteristics: a central location with a population from which to attract audiences; adequate performance and accommodation space; and the availability and size of local subsidies. Beyond this, the festival was a "destination in itself," and the specific city or town was not important.

Despite the agreement on the importance of place for festivals, however, a number of authors note the paucity of research on the geography of festivals (Visser 2005, 155; Van Aalst and Van Melik 2012, 196; Cudny 2014, 135). To address this, Cudny (2014, 135) listed a number of research themes that could be explored in geographical festival research. These included the role of place in cultural creation and consumption; a social analysis of festivalgoers (both local and non-local) in terms of their origins, motivations, and attitudes; the economic impact of festivals on the host economy; as well as consideration of festival typographies and aims. A study of the geography of festivals at a local level could also include available infrastructure (physical space), social spaces (non-physical space), social flows and interactions, cultural space, and political space. Cudny ties the place identity role of festivals to "image space," which includes ways in which the event is perceived by tourists and local residents and the media (138).

Festivals can have a number of potential impacts on their host towns and attendees. One of the most commonly measured effects is the economic (or financial) impact that festivals have as a result of the additional spending of tourists who participate in the event (Crompton, Lee, and Schuster 2001, 81; Snowball and Antrobus 2002, 1299; Snowball 2008, 66; Snowball and Seaman 2017, 31). Since spending by local residents is not normally included in such calculations, the proportion of tourists who can be attracted from outside the host city is an important determinant of the resulting economic benefits. As Saayman and Saayman (2006, 581) demonstrate, the location of festivals—whether close to or far away from—large metropolitan areas from which non-local audiences can be drawn is thus also a significant factor in determining economic impact.

However, festivals can also have other, equally important social and cultural impacts. As Quinn (2010) points out, arts festivals have their own intrinsic cultural goals related to things like celebrating a particular genre, developing audience appreciation of an art form, and encouraging risk-taking and innovation among performers. Cultural activities can also promote the development of social capital and a sense of communal solidarity, identity, and pride among local residents.

Quinn (2006) studied the non-economic impacts of festivals in Ireland, which mostly emerged as small, local initiatives staffed by volunteers. In Ireland, hosting a festival is generally associated with the development of the arts in the cities in which they are held; it is also associated with greater cultural participation, better cultural infrastructure, and increased year-round cultural activities. Festival plans that focus too much on the economic impact of the event to the exclusion of other values are unlikely to be sustainable, as Ivanovic (2008) points out, since the festival is at risk of losing its cultural meaning and uniqueness.

A festival's geography can be an important determinant of the uniqueness and authenticity of its cultural offerings, not to mention the socio-economic benefits derived by the host economy. However, despite some attempts to develop frameworks for the geographical analysis of festivals (Ma and Lew 2012, 21; Cudny 2014, 139), this area of research has not received much attention, either internationally or in South Africa. This chapter, in addition to a traditional economic impact analysis of the event, sets out to analyze which aspects of geography affect the process of forging place identity via the Mahika Mahikeng Festival.

Context

A BRIEF OVERVIEW OF THE TOWN

Mahikeng has a complex history due to the intertwining of Batswana, Boer, and British cultures and powers. Mahikeng has long been something other than an ordinary South African town and its main claim to international renown is the siege of Mafeking (as the town was then known[1]) by Boer forces during the Anglo-Boer War. The first settlement, however, was founded by the Tshidi Rolong on the banks of the Molopo River. The African settlement was named Mahikeng, meaning "the place of stones." Being on the imperial road to the north from Cape Town and Kimberley to Rhodesia, the British sought to secure the area and so established a colonial town adjacent to the African settlement, which they named Mafeking; the town would henceforth serve as the extraterritorial capital of the Bechuanaland Protectorate. This brought a British cultural influence to the area that was unique for its time as the surrounding settlements were primarily Afrikaans. The British cultural influence has remained to this day: Mahikeng is still a primarily English-speaking town with several prominent heritage buildings that hark back to its colonial past. This includes the Mahikeng Museum, which had displays on the siege and the Boy Scouts (founded during the siege) for many years. By the early 1960s, Mafeking's main economic activities revolved around its status as a colonial capital, agriculture and agricultural processing, railways, as well as commerce and trading.

This prosperity did not last. In 1966, the Bechuanaland Protectorate gained independence and the capital was moved to Gaborone, in the newly independent Botswana. Up until this time, Mafeking had managed to remain somewhat apart from apartheid oppression as petty apartheid could not be enforced as stringently as elsewhere due to the British colonial presence. However, the changing political situation left Mafeking to experience the full impact of apartheid segregation and economic malaise (Parnell 1986, 205).

Under the policy of grand apartheid, the Bophuthatswana Bantustan (the so-called "homeland" of the Batswana people) was created and granted "independence" from South Africa in 1977. Uniquely for South African towns, the white residents of Mafeking voted to leave the jurisdiction of South Africa in 1980 and join the African-ruled Bantustan. The motivation for this was largely economic as business owners and residents wanted to recover the town's status as a capital, with the accompanying economic benefits, and to take advantage of the lower taxes levied within Bophuthatswana. This meant that Batswana and British cultural heritage was preserved as Afrikaaner nationalism could not take hold within the town. The town was renamed Mafikeng at this time, with the change in spelling signalling a move back toward the Setswana language, and from 1980 to 1994, the town was something of a "bantustan boomtown" (Drummond and Parnell 1991, 167). It benefited once again from being a political capital and was heavily subsidized by the South African state. It also experienced an infrastructure construction boom in tangible cultural assets, though this was not a deliberate LED policy, and it benefited from mining royalties and tourism spending (Drummond 2019). These tangible cultural assets formed the basis for a cultural and creative industries cluster in the town that included a local newspaper, radio station, and television broadcaster; orchestra, dance, drama, and arts and crafts at Mmabana Cultural Centre; recording studios (where the soundtrack to the Lion King was produced); Lotlamoreng Cultural Village; 60s Festival; and the University of Bophuthatswana's fine art department, which ran the Sol Plaatjie Memorial Art Exhibition (Drummond 2019). The intangible cultural assets illustrate the combination of British and Batswana cultures as the two co-existed through the arts, with classical music and ballets being performed alongside dramas written and produced by Batswana creatives relating to their culture and the situation in South Africa (Drummond 2019).

With democracy looming, the Bantustans were disbanded and Mafikeng was incorporated into the new South Africa. At this time, there was considerable anxiety regarding the future of the town among investors, whose continued confidence depended on Mafikeng's ability to retain its status as the provincial capital of the new North West province, which was established in 1994. Following the end of apartheid and the expansion of democracy, the new African National Congress (ANC) government set about dismantling the legacies of the Bantustans, which were viewed as institutions of apartheid. Many cultural assets were consequently lost as Mafikeng's orchestra was disbanded and its television and recording studios were shut down. In terms of tangible and intangible culture, the town is a shadow of its former self: it is no longer a cultural melting pot, nor does it have the cultural infrastructure and institutions to support a vibrant cultural economy.

Indeed, Mafikeng has been plagued by insecurity in the post-apartheid period, which saw the town generate continued negative sentiment resulting in an effective investment strike. Therefore, while much of South Africa prospered, Mafikeng suffered. Continued uncertainty over its status as a provincial capital cast a shadow over the town's economy, with the result that jobs were lost and the property market floundered. However, the new ANC government strategically decided to leave the capital in Mafikeng and, after the 2004 elections, renewed its commitment to supporting growth in the town. Confidence swelled and resulted in another boom phase for the town driven by investment and construction.

More recently, the town has once again entered into a slump due to the after-effects of the 2008–9 financial crisis, along with the failure to promote development and government incapacity. This is illustrated by the attempt to establish the Mafikeng Industrial Development Zone (MIDZ), which was based around the local airport but failed to take off. The rationale for the MIDZ iniative was to promote the town as an investment destination centred around an aviation hub, which would import component parts and export finished value-added products. Though considerable infrastructural developments were delivered at the site of the MIDZ, the success of the scheme required national-level political support and coordination with the Department of Trade and Industry. The lack of national buy-in is largely responsible for the scheme's failure. The most recent developmental attempt has been the rebranding of Mahikeng (renamed in 2012 in a return to the town's Batswana roots) as the cultural capital of South Africa (Nel and Drummond 2017, 10). This is based on place marketing and focuses on the Mahika Mahikeng Festival, which focuses on celebrating the Setswana language and as such is illustrative of a resurgence in Batswana culture.

Festivals in Mahikeng

A festival, held each year at the same time, has been in place in Mahikeng for roughly twenty years. Earlier versions included the 60s Festival, which ran for sixteen years on a farm about 20 kilometres outside of the town. This was regarded as a "rite of passage" festival for local young people; in this it was similar to other festival experiences in South Africa geared toward school-leavers (Rogerson and Harmer 2015, 226). The event was aimed at African partygoers from Mafikeng, Gaborone, Maseru, and Johannesburg, and it explicitly targeted a young audience, many of whom were outsiders. It thus promoted the phenomenon of cultural experience tourism (Richards 2011, 1228), attracting tourists interested in absorbing the knowledge of endogenous cultural assets. However, there was significant local opposition to this festival as many people thought it promoted public drunkenness, promiscuity, and dangerous driving, resulting in fatalities. The 60s Festival thus died out and was replaced by the Legends of House Festival, which ran in the town for three years. This was organized by a local music producer and DJ and was held at the same time of year. The Legends of House Festival attracted top performers from the United Kingdom and the United States, and it was initially highly successful as these top performers attracted large numbers of festival tourists. Despite this, when it faced competition from the newly established Mahika Mahikeng Festival, a local government-sponsored event, the Legends of House organizer chose not to offer it against this government-favoured alternative.

The latest festival to take place in the town is the Mahika Mahikeng Music and Cultural Festival, which occurs over four days in early December (8–11 December in 2016). Although it takes place at a similar time as the older events, it began in its current form in 2015. It is described by the organizers as "A music and cultural festival, which aims to provide a platform for product positioning and paradigm innovation for the creative industry sector in Bokone Bophirima and the development of Mahikeng as the capital of the Arts in the country" (Snowball and Antrobus 2017, 5). It is thus a government LED initiative that seeks to capitalize on the development potential of the cultural and creative industries and cultural tourism and is linked to what is known as the "Mahikeng Rebranding, Repositioning & Renewal Programme" of the provincial government. As such, the festival's goals include promoting cultural and heritage tourism; celebrating artists in the region and across the nation; repositioning and rebranding Mahikeng and the North West province as a cultural hub;

stimulating economic growth; and creating jobs in the music and cultural industries. These goals, along with the festival's links to the provincial rejuvenation program, align with the cultural turn observed in international policy and development strategy. Richards (2011, 1231) describes how festivals and other cultural and creative tourism events are being used to boost economic activity through spillovers, achieve social and cultural outcomes, and add value to place through branding. By creating the Mahika Mahikeng Festival, the provincial government is attempting to buy in to this culture-led development approach, as the festival represents the latest policy initiative to promote LED in Mahikeng. As Tomaz explains in chapter 2 of this volume, though it is more commonly employed in large cities, Tourism has become a popular culture-led development strategy, one that exploits the particular local symbolic, socio-cultural, relational, and territorial assets of the place. In this case, the festival is attempting to use cultural tourism and the cultural and creative industries to support placemaking processes that recreate the town's imaginaries and contribute to the revitalization of the town; the diversification of tourism offers (though not tourists' profiles, as the festival was intended to mainly appeal to Batswana); the development of the cultural and creative sectors, including the conservation and preservation of material and immaterial Batswana heritage; and the promotion of social cohesion among the prominent local Batswana culture.

The festival includes many different features, among them an arts and crafts market (free entrance) and music concerts featuring well-known and emerging South African artists performing jazz, traditional and contemporary dance, drama, and gospel music. It markets itself as a celebration of the Setswana language and the Batswana culture, a creative representation of place, and so seeks to achieve a national objective of promoting social cohesion through the reinvigoration of local culture for both locals and tourists. For the latter, the festival's attraction lies in the authentic cultural experience that it and events like it offer (Richards 2011, 1237). In an attempt to be more inclusive, the Department of Culture, Arts and Traditional Affairs called all citizens of the province to give input on how the festival should be run and organized. This intention certainly speaks to the collaborative aspect of this volume's operating definition of creative tourism; however, it seems the call did not receive the response that organizers were hoping for. A post on the festival's Facebook account in September 2016 indicated that "less than 10 submissions were received" (quoted in Snowball and Antrobus 2017, 5). According to the festival's CEO, artists were also chosen with the help of municipal districts, which nominated performers who had, for example, won regional talent competitions, as well as suggestions from social media platforms. However, the attempt to be collaborative and involve locals in the organization of the festival was not successful. A protest was staged by the South African Arts and Culture Youth Forum on the first day of the festival (8 December 2016) against what they saw as the deliberate exclusion of some emerging artists from the program. Moreover, though this was not evident at the time of the 2016 festival, allegations have since emerged that the failure of local artists and service providers and government organizers to collaborate was due to the government's decision to award tenders and performing contracts to non-local service providers and artists. This of course undermined what was supposed to be a festival celebrating the local Batswana culture and Setswana language (Mashigo 2019).

Research Methods

To measure the impact of the Mahika Mahikeng Festival, and thus its touristic and developmental contributions, the festival must be valued. The values associated with arts, culture, and heritage can be divided into three broad categories: economic (financial) impacts, social impacts, and the intrinsic value of art itself. Economic impacts come about as a result of the inflow of new money into an economic system as a result of visitors from outside the region—in this case, tourists to the Mahika Mahikeng Festival. Visitors spend on accommodation, transport, food, shopping, tickets, etc. This spending then recirculates in the host economy, increasing sales and employment in local businesses. Social values relate to the benefits to society, such as education, creativity and innovation, social cohesion, and identity formation (Bohm and Land 2009, 76). Intrinsic values are related to the symbolic, artistic nature of the product itself and to feelings invoked in individual participants (such as joy, sadness, anger, delight, and curiosity). All three of these values contribute to placemaking and sustainable cultural development. The first two are measurable and tangible; the third is intangible but entwined with the collaborative relationship between creative producer and consumer.

Based on these three broad categories of value, the *M&E Framework* (SACO 2016, 32) outlines the development of five cultural value indicators or themes: audience development and education; human capital and professional capacity building; inclusive economic growth; social cohesion and community development; and reflective and engaged citizens.

The value indicators all feed into the conditions in host communities, and the type of cultural events, that are more likely to lead to sustainable cultural development. For example, festivals that have a strong sense of place and connect to local cultures and identities are more likely to build social cohesion and community development. Those that involve local organizers and artists contribute to the development of human capital (skills and experiences), and they are more attractive to cultural tourists as "authentic" experiences, thus better enabling the monetization of local cultural assets and increasing economic impact. As will be demonstrated in our case study of the Mahika Mahikeng Festival, the lack of local involvement (of both suppliers and artists) resulted in few of the benefits (social, cultural, or economic) being realized, and the subsequent collapse of the festival.

However, each cultural event is different, and any useful valuation study needs to take into account the stated aims of the project/event/organization, what the expected impacts are, and who is expected to benefit. The monitoring and evaluation tool would then have to be designed based on the answers to these questions. Based on the contextual information about the Mahika Mahikeng Festival and the *M&E Framework*, the following values, indicators, and data-gathering methods were identified (see table 3.2).

Indicators that allow us to gauge the success of the event in terms of culture-driven local economic development are those relating to the ability of the festival to attract visitors (in the audience development and education theme identified below), the involvement of local artists and suppliers, and marketing and place identity (social cohesion and community development theme), and the economic impact of the event (inclusive economic growth theme).

Table 3.2. Value Themes, Indicators, and Research Methods at the Mahika Mahikeng Festival 2016

THEME	EXAMPLES OF INDICATORS	RESEARCH METHOD
Audience development & education	Demographics (age groups; cultural/racial groups; gender)	Audience survey
	Origins (local/tourist; rural/urban; province; nationality)	
	Income and education groups	
	Participation and time use	
	Ticket sales/participation (numbers)	Organizer data (Computicket data and crowd size estimation)
	Building cultural capacity	Audience survey (opinions)
Human capital/ professional capacity building	Experience gained by local, emerging artists	Organizer data on performers, procurement policy, and/or use of local service providers, etc.
	Showcasing South African art and artists	
Inclusive economic growth	Organizer spending	Organizer data
	Earned income/turnover	
	Sponsorship (and sources)	
	Audience spending	Audience Survey
	Length of stay (bed-nights)	
	Tourist leverage (extended trips)	
	Contribution to gross geographic product/gross value added	Economic impact analysis
Social cohesion & community development	Artist/producer origins (diversity)	Organizer data
	Cultural offerings (local arts/heritage/languages)	Audience survey
	Audience diversity	
	Audience opinions	
	Marketing and place-identity	
Reflective & engaged citizens	Appreciation of diversity	Audience survey
	Fostering dialogue and intercultural understanding	
	Developing pride in local cultures	

This research made use of two main data sources:

1. An audience survey administered at the various festival events by trained enumerators. The questionnaire measured audience demographics, opinions on the values the event creates, the role of the festival in rebranding Mahikeng, self- reported behaviours (for example, the parts of the festival attended), and, for non-local visitors, spending as a result of the event.

2. Data that can be obtained from festival organizers. For example, Mahika Mahikeng uses Computicket as its ticket vendor for some of its larger music events. Computicket can provide information not only on the number of tickets sold, but also on where and when they were sold; this information can be very useful in determining, for example, the origins of festivalgoers. Organizers can also provide data on sponsorship, direct employment, partnerships with local service providers, and the diversity of events and artists on offer.

This data was collected from festivalgoers by a team of four trained interviewers and through key stakeholder interviews conducted by research leaders, as well as through observation of, and attendance at, as many of the festival activities as possible. A total of 380 interviews were conducted with as wide a variety of attendees as possible. Furthermore, the information on the origins of festivalgoers was mapped using various geographic information systems. For example, a heat map was created to show where attendees were from; as tourist numbers affect the economic impact of the festival, this data helps us gauge the festival's success at attracting creative tourists to a creative arts festival. Attendees were mapped from the nine provinces of South Africa, the Ngaka Modiri Molema Municipality in which Mahikeng is located, and the town itself, with darker colours representing places from which more festivalgoers originated (see figure 3.1).

Table 3.3. Summary of Mahika Mahikeng Survey Results

THEME	INDICATORS	SUMMARY OF RESULTS
Audience development & education	Demographics (age groups; cultural/race groups; gender)	78% Setswana; 8% Sesotho; 6% isiXhosa speakers; 59% men; 80% under 36 years old
	Origins (local/visitor; rural/urban; province; nationality)	65% locals; 98% South Africa; 1.3% Botswana
	Income and education groups	54% have household income less than R10,000 p/m; 19% more than R20,000 p/m; 61% tertiary education
	Participation & time use	57% attended Jazz; 37% Motswako; 50% attended 1 event; 27% attended 2; average 1.53 tickets per person
	Ticket sales/participation (numbers)	Estimated total attendance 3,500; estimated total ticket sales 5,355
	Building cultural capital	62% agree that "the festival increases my understanding of African cultures" (20% neutral)
Human capital/ professional capacity building	Experience gained by local, emerging artists	Call for inputs from province; call for nominations from municipal districts and on social media; mix of local and national artists (some contention)
	Showcasing South African art and artists	Wide variety of genres included presented by local and national artists

Table 3.3. (*continued*)

THEME	INDICATORS	SUMMARY OF RESULTS
Inclusive economic growth	Organizer spending	Pre-event cost estimate of R12.9 million; estimated spending in impact area, funded by non-local sources R4.2 million
	Earned income/turnover	Ticket sales pre-event estimate R2 million
	Sponsorship (and sources)	R2 million DAC; other sponsors included: Culture, Arts and Traditional Affairs (R6 million); Mahikeng Rebranding, Reposition, and Renewal Program; Mmbana, North West Provincial Government; ABSA; National Lottery
	Audience spending	Visitor average spending of R832 per person; R619 for local residents
	Length of stay (Bed nights)	1.5 nights and 2.4 days; 45% day visitors
	Tourist leverage (extended trips)	4.3% "staying on in the region to visit tourist places nearby"
	Contribution to Gross Geographic Product/Gross value added	Estimated at R7,05m impact on Mahikeng and surrounds.

Table 3.3. (*continued*)

THEME	INDICATORS	SUMMARY OF RESULTS
Social cohesion & community development	Artist/producer origins (diversity)	Mix of NW Province and national (some contention about how local artists chosen)
	Cultural offerings (local arts/ heritage/languages)	Wide variety of genres in mix of languages
	Audience diversity	65% locals; 98% South Africa; 1.3% Botswana; 78% Setswana; 8% Sesotho; 6% isiXhosa speakers; 59% men; 80% under 36 years old; 61% tertiary education
	Marketing & place identity	32% strongly agree and 45% agree that "the festival is part of what makes Mahikeng a special place"
Reflective & engaged citizens	Fostering dialogue and intercultural understanding	57% agree and 20% strongly agree that "the festival is an event where people from different cultures and backgrounds can meet and talk together"
	Developing pride in local cultures	77% agree that "festival makes me feel proud of my cultural heritage"

*Organizer data was provided in advance of the event (based on projections), but the research team was not able to obtain post-event data. Figures for organizer spending, the proportion of local sponsorship, ticket sales, and attendees are thus based on the observations of the research team and/or the data provided by organizers in advance.

Results and Discussion

The survey and organizer data was used to determine the festival's success as an LED initiative through an analysis of whether or not it achieved its goals of promoting cultural and creative tourism, celebrating local and national artists, rebranding Mahikeng as a cultural capital, and stimulating the local economy. The results of the survey are presented in table 3.3.

Promoting Cultural Tourism, Social Impacts, and Place Identity

Firstly, in terms of promoting cultural tourism, festival attendees' origins need to be analyzed. The map in figure 3.1 shows festivalgoers' origins, with darker colours representing areas of higher attendance. According to survey results, almost two-thirds (65 per cent) described themselves as "local Mahikeng residents." The remaining 35 per cent were visitors, but, as the map shows, the majority of tourists were from areas closer to Mahikeng. The festival attendance numbers appear to radiate outwards as, after Mahikeng itself, the largest groups of festivalgoers were from the local municipality, the North West province, and neighbouring provinces. Furthermore, the vast majority (98 per cent) of attendees were South African, with only a small number from Botswana (1.3 per cent), Lesotho (0.3 per cent), and Namibia (0.3 per cent). This is significant as it suggests that the appeal of a festival that celebrates one particular culture is highly localized to areas where Setswana is spoken widely and where the Batswana culture and people are found. This is also supported by survey results as 87 per cent of visitors reported that the festival was their main or only reason for coming to Mahikeng. Moreover, while there was some diversity in terms of cultural groups (using home language as a proxy), more than three-quarters (78 per cent) of attendees spoke Setswana at home. This suggests that, while there is a demand for a Batswana cultural festival, it is expressed primarily by those who are connected to this culture. This in itself is a success as it links to the goals of rejuvenating Batswana culture, instilling pride in local people, and promoting social cohesion.

These social impacts are supported by the opinion questions in the survey, which are designed to gauge the festival's non-market impacts in such areas as building cultural capital and social cohesion. Attendees were asked to respond to statements designed to measure this with responses ranging from strongly agree to strongly disagree. To the statement "The festival makes me feel proud of my cultural heritage," 41 per cent of festival attendees agreed and 36 per cent strongly agreed. However, to the statement "The festival increases my understanding of African cultures," there was a somewhat more negative response. While the majority (62 per cent) had a positive view of the festival's role in building their cultural capital, most of these were in the "Agreed" category (51 per cent), while only 11 per cent strongly agreed. Additionally, local buy-in is crucial to the success of new cultural events that require local support while they are still small and finding their feet. Overall, cultural tourism promotion was limited as attendees were mainly from Mahikeng (see figure 3.1). This will in turn have an effect on the event's economic impact since locals do not add to spending and visitors from nearby spend less since it is easy for them to return home after attending a festival event. Consequently, in terms of promoting cultural tourism, the festival was a failure as the government organizers took a top-down approach, which locals felt excluded their input, services, and performers. This resulted in poor buy-in from locals as well as an anti-festival movement. There were also poor tourist numbers, due mainly to missed marketing opportunities, which could have targeted Batswana communities in nearby Botswana and in the rest of South Africa, thus supporting the festival's goals of celebrating the Setswana language, the Batswana culture, and promoting social cohesion among this community.

One of the festival organizers' important stated aims is to use the event for the "development of Mahikeng as the capital of the Arts in the country." While the festival is still in its early days, attendees were asked to respond to the following statement (on the "Strongly Disagree" to "Strongly Agree" scale) to provide a baseline measure of effectiveness: "The festival is part of what makes Mahikeng a special place." Results show that the festival is already regarded by the majority (77 per cent) of attendees as being an important part of place identity in Mahikeng (with 32 per cent strongly agreeing, and 45 per cent agreeing).

Figure 3.1. Origins of Mahika Mahikeng Festival Attendees

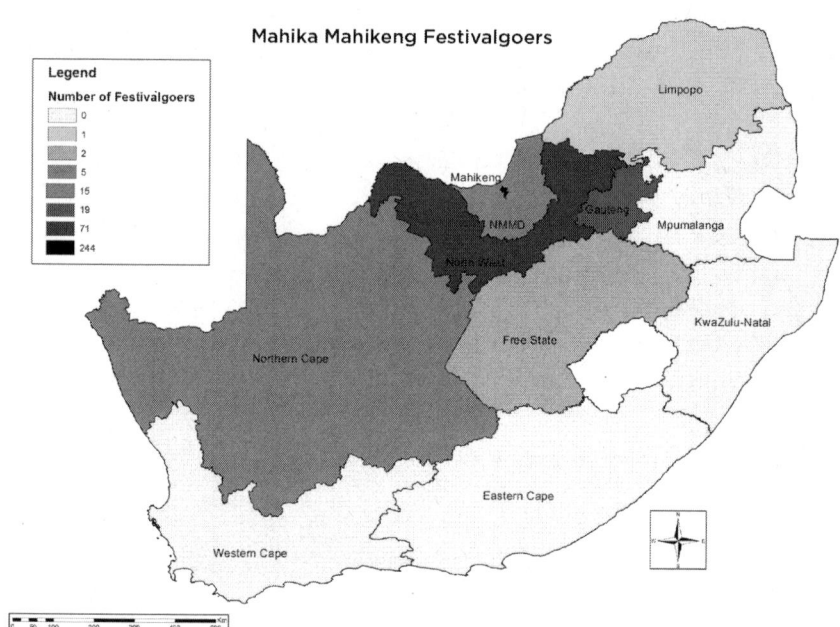

The Economic Impact of Mahika Mahikeng on the Host Economy

Festivals can contribute quite significantly to the economy of their host town by improving the location's image or brand, as well as by attracting cultural tourists. In a South African study, Toerin (2020) found that smaller towns especially are dependent on the tourism sector for much of their economic activity, and that towns with more tourist infrastructure are associated with lower levels of poverty. Nadotti and Vannoni (2019) also note that economic impact studies provide a powerful way to quantify and measure the impact of a festival on a regional economy.

Economic impact starts with the first round, or direct impact, of spending by visitors and festival organizers, followed by indirect and induced expenditure as the initial injection is re-spent (known as the multiplier effect). The economic geography of the host town or city affects the size of the economic impact in two ways. Firstly, the economic structure of the city determines the size of the multiplier effects, as well as the level of "leakages"—the outflow of money as event organizers and local suppliers and consumers buy things from other regions. The smaller the area, the greater the amount of re-spending that takes place outside of it, and the smaller the multiplier size (Seaman and Snowball 2017, 5). Secondly, the distance of the host from other cities from which cultural tourists can be attracted (as well as, of course, the effective marketing of the event) affects the number of visitors to the festival.

Since input-output tables are not generally available in South Africa at the municipal or city level, multipliers can be estimated by looking at past studies and regional characteristics. An economic impact calculator, developed for the Georgia Department of Economic Development and adjusted for South Africa with the assistance of the original modeller (Seaman and Snowball 2017, 31), was used to estimate the multiplier in this study.

Based on data from Statistics South Africa (2018), the population contained within the festival impact area, defined as "Mahikeng and surrounds," was estimated at 250,000, which has an estimated expenditure multiplier of 1.42. This means that, for every R1 of "new" spending in the economy, a further 42c is generated in indirect and induced impacts.

The total number of festival attendees, both local residents and visitors, was estimated using pre-event organizer estimates and crowd counts at the events at which interviews were conducted. Taking into account that the average number of shows/events attended by each person was 1.53 (from visitor survey data), it is estimated that 3,500 people attended the festival. This implies that 5,355 tickets were sold (3,500 x 1.53), which is within the range of what organizers anticipated (5,000–10,000).

Based on interview data and previous studies, it was estimated that 55 per cent of festivalgoers were local residents, while 45 per cent (1,575) were visitors. However, some of these visitors were not in the area specifically or mainly to attend the festival (13 per cent) and are thus likely to have come to the city even if the event had not taken place. The total number of non-local visitors in the area specifically to attend the festival was estimated to be 1,370.

Average spending per visitor for the whole trip was R832 (data from the visitor survey). Gross visitor spending was thus R1.14 million, and net visitor spending (allowing for immediate leakages) was R775,000. Including organizer spending sourced from outside the host city that was spent locally (R4.2 million), as well as multiplier effects, the total economic impact of the 2016 Mahika Mahikeng Festival on the economy of Mahikeng of just over R7 million.

This very modest financial impact (given that organizers estimated the cost of hosting the event at nearly R13 million) is clearly the result of the low numbers of tourists attracted to the festival from outside of the host city, and the small-town nature of Mahikeng's economy. It could be offset by the positive non-market experiences of those who did attend, such as the 62 per cent who reported that the festival increased their understanding of African cultures, which could lead to further appreciation of, and participation in, cultural events among some attendees in the future. Audiences were also relatively diverse in terms of gender and language, and most (80 per cent) were young.

Efforts were made to include performers from as wide a variety of local and national settings as possible, which meant that the festival can play a role in building the reputations and experiences of performers. However, some local artists staged a protest on the first day of the festival, saying that they had been deliberately excluded. Local suppliers of other tourist infrastructure and equipment also claimed that they had been passed over by organizers in favour of more politically connected businesses in Johannesburg.

Festival sustainability is crucially dependent on the co-operation and goodwill of the local community (Quinn, 2006), and on the perceived authenticity and cultural value of the event (Ivanovic 2008). Since our survey at the 2016 Mahika Mahikeng Festival, the hosting of the event became increasingly contentious, with almost all the focus being on which companies would be granted the lucrative tenders to run the event, and not on its supposed social or cultural aims. In 2019, there were media reports that disagreements over festival tenders were threatening to destabilize the whole province (Mashigo 2019), and in the end, the festival was not held. It remains to be seen whether, as the organizers intend, it will be held again in 2020.

Conclusions and Recommendations

The Mahika Mahikeng Music and Cultural Festival took place from 8–11 December 2016. While the festival included a variety of genres and a mix of local, provincial, and national performers, it was nonetheless criticized by attendees for having little diversity as it adhered closely to the Batswana culture theme. The event's main aims were to promote cultural and heritage tourism; to celebrate artists in the region and nation (with a focus on musicians from the province); to reposition and rebrand Mahikeng and the North West province as a cultural hub; and to stimulate economic growth and create jobs in the music industry. As shown in table 3.3, the majority of the audience were local residents whose home language was Setswana (78 per cent). The festival's target audience is thus relatively narrow, and since there were not many tourists, visitor spending was limited, which circumscribed the festival's stated aims of promoting LED and tourism. However, the majority of respondents (62 per cent) agreed that "the festival increased my understanding of African cultures." In the audience development category, the festival thus performed quite well, although it does not appear to have been successful in attracting audiences from other neighbouring countries, especially Botswana and Namibia, which was one of the organizers' aims.

An area that could be improved is marketing and information, especially when it comes to attracting tourists to the festival. Respondents commented that it was difficult to obtain information about the festival in advance. Approximately ten days before the event, local marketing began in earnest, with posters, radio, and newspaper advertisements in and around Mahikeng itself, but this did not extend further afield and so the effect was primarily local. The festival also had a fairly active social media presence (Facebook and Twitter). As noted by some of the respondents, festival organization (in terms of things like information provided on starting times and venues, the names of performers for specific events, etc.) also needs to be improved. Some performers felt that the lack of accurate information and marketing led to small audience sizes.

The number of tickets sold was estimated at 5,355 (based on an estimate of 3,500 attendees buying an average of 1.53 tickets each). The festival's economic impact on Mahikeng and the surrounding area was estimated at just over R7 million. This did not represent a significant return on investment and so the economic contribution of the event was limited. To increase economic impact in the future, and to allow the festival to play a greater role in the "development of Mahikeng as the capital of the arts in the country," a greater percentage of non-local visitors should be encouraged, especially visitors from outside the North West province, who are more likely to stay overnight and visit other places in the area. However, the festival is already gaining some recognition as a builder of place identity: 77 per cent of attendees agreed that it was part of "what makes Mahikeng a special place."

The festival also provided a platform for local and provincial artists to showcase their work across a variety of genres. Selection included asking for nominations from municipal districts and taking recommendations from social media platforms. However, there was still some dissatisfaction, with some local groups feeling that they had been deliberately left out and that more local artists (or even exclusively local artists) should have been included. The balance between less well-known local artists and national or international performers is always a challenge for festival organizers because including too high a proportion of local artists can reduce audience numbers and the ticket prices that could be charged. A more transparent selection process may be part of the solution. Furthermore, given that the festival is mainly attended by locals, it would be beneficial to collaborate more with local service providers and artists and to engage with the local community in a more inclusive grassroots approach to ensure that the festival is appealing to its intended audience and is more sustainable.

There is considerable potential to expand the festival. An unusual feature in a medium-sized town, Mahikeng has a wide variety of large, high-quality performance spaces already equipped with seating, lighting, and sometimes sound equipment. The availability of these venues gives Mahikeng a comparative advantage over other similarly sized towns and provides an opportunity for expanding the festival as, unlike in other medium-sized towns, Mahikeng would not be constrained by supply-side limitations. If the festival were to grow, accommodation offerings could also be expanded by using residences at Mafikeng campus of the North-West University and hostels at, for example, the International School of South Africa, both of which are on holiday by the time of the main festival weekend. Both the university and the school also have large, well-equipped auditoriums. The festival offerings can also be improved as suggestions from interviewees (mainly locals) included having some activities for children, using malls and government buildings as exhibition spaces, and having food and drinks available for purchase where arts and crafts are sold. The jazz and Motswako events could be expanded as they were the most popular, and moving the craft market to a more central and accessible location would attract more business. Furthermore, there is a suggestion that organizers should consider moving the festival dates, which as mentioned overlap with student holidays, to maximize attendance by university students who live in the town. These suggestions from interviewees show that there is a desire among locals to influence the festival's organization, and a belief that a more collaborative approach could be fruitful. If the festival were to broaden its scope beyond the promotion of Batswana cultural heritage and also target a younger audience, it would be likely to attract more out-of-town festivalgoers. This would be akin to bringing in more tourists with a higher spending impact. The town has the potential to fulfill the aim of rebranding and placemaking around creativity and tourism, but better festival organization and co-operation with other stakeholders has to be advanced.

NOTES

The authors wish to acknowledge the financial support provided for this research by the South African Cultural Observatory, a research organization funded by the South African Department of Sport, Arts and Culture.

1   The spelling of the town's name has changed over time depending on who was in power.

References

Bohm, Steffen, and Chris Land. 2009. "No Measure for Culture? Value in the New Economy." *Capital & Class* 33, no. 1 (March): 75–98. https://doi.org/10.1177/030981680909700105.

Crompton, John L. 2006. "Economic Impact Studies: Instruments for Political Shenanigans?" *Journal of Travel Research* 45, no. 1 (August): 67–82. https://doi.org/10.1177/0047287506288870.

Crompton, John L., Seokho Lee, and Thomas J. Schuster. 2001. "A Guide for Undertaking Economic Impact Studies: The Springfest Example." *Journal of Travel Research* 40, no. 1 (August): 79–87.

Cudny, Waldemar. 2014. "Festivals as a Subject for Geographical Research." *Geografisk Tidsskrift-Danish Journal of Geography* 114, no. 2 (March): 132–42. https://doi.org/ 10.1080/00167223.2014.895673.

Drummond, James H. 2019. "The Historical Evolution of the Cultural and Creative Economy in the North West Province, South Africa: Implications for Contemporary Policy?" Paper presented at the Annual International Conference of the Royal Geographical Society, Imperial College London, 30 August 2019.

Drummond, James H., and Susan M. Parnell. 1991. "Mafikeng-Mmabatho." In *Homes Apart: South Africa's Segregated Cities*, edited by Anthony Lemon, 162–73. London: Paul Chapman.

Duxbury, Nancy, and Heather Campbell. 2011. "Developing and Revitalizing Rural Communities through Arts and Culture." *Small Cities Imprint* 3, no. 1 (Spring): 111–22.

Ivanovic, Milena. 2008. *Cultural Tourism*. Cape Town, South Africa: Juta & Company.

Kavese, Kambale. 2012. "Eastern Cape Automotive Sector Analysis: An Economic Model for Policy and Investment Development." Eastern Cape Socio Economic Consultative Counsel. http://www.iioa.org/conferences/20th/papers/ files/994_20120320030_ECAutomotivesectoranalysisfinaldraft20March2012.pdf.

Ma, Ling, and Alan A. Lew. 2012. "Historical and Geographical Context in Festival Tourism Development." *Journal of Heritage Tourism* 7, no. 1 (Spring): 13–31. https://doi.org/10.1080/1743873X.2011.611595.

Mahika Mahikeng. n.d. "About." Accessed 15 April 2018. https://www.facebook.com/pg/ MahikaMahikeng/about/?ref=page_internal.

Mashigo, Lehlononolo. 2019. "Mahika Mahikeng Tender Row Threatens to Destabilise the North-West Province." *The Star* (Johannesburg), 6 November. https://www.iol. co.za/the-star/news/mahika-mahikeng-tender-row-threatens-to-destabilise-north-west-province-36746078.

Nadotti, Loris, and Valeria Vannoni. 2019. "Cultural and Event Tourism: An Interpretive Key for Impact Assessment." *Eastern Journal of European Studies* 10, no. 1: 115–31.

Nel, Verna, and James H. Drummond. 2017. "From Mafikeng to Mahikeng: Spatial Transformation in a South African Provincial Capital." South African Cities Network. Accessed 15 April 2018. http://www.sacities.net/wp-content/ uploads/2017/10/Mahikeng.pdf.

Parnell, Susan M. 1986. "From Mafeking to Mafikeng: The Transformation of a South African Town." *Geojournal* 12, no. 2 (March): 203–10.

Quinn, Bernadette. 2006. "Problematising 'Festival Tourism': Arts Festivals and Sustainable Development in Ireland." *Journal of Sustainable Tourism* 14, no. 3: 288–306.

———. 2010. "Arts Festivals, Urban Tourism and Cultural Policy." *Journal of Policy Research in Tourism, Leisure and Events* 2, no. 3: 264–79.

Richards, Greg. 2011. "Creativity and Tourism: The State of the Art." *Annals of Tourism Research* 38, no. 4 (October): 1225–53. https://doi.org/10.1016/j.annals.2011.07.008.

Rogerson, Jayne M., and Devin Harmer. 2015: "A 'Rite of Passage' Youth Festival in South Africa: the Origins, Attendees and Organization of Matric Vac." *Nordic Journal of African Studies* 24, nos. 3–4: 221–40.

Saayman, Melville, and Andrea Saayman. 2006. "Does the Location of Arts Festivals Matter for Economic Impact?" *Papers in Regional Science* 85, no. 4 (November): 569–84. https://doi.org/10.1111/j.1435-5957.2006.00094.x.

SACO [South African Cultural Observatory]. 2016. *A Framework for the Monitoring and Evaluation of Publicly Funded Arts, Cultural and Heritage.* Port Elizabeth: South African Cultural Observatory.

Snowball, Jeanette. 2008. *Measuring the Value of Culture: Methods and Examples in Cultural Economics.* Berlin: Springer-Verlag.

Snowball, Jeanette, and Geoff Antrobus. 2002. "Valuing the Arts: Pitfalls in Economic Impact Studies of Arts Festivals." *South African Journal of Economics* 70, no. 8 (December): 1297–1319.

———. 2017. "South African Cultural Observatory Monitoring and Evaluation: Key Development Indicator Report on DAC Interventions Intervention: Mahika Mahikeng Festival." South African Cultural Observatory. https://www.southafricanculturalobservatory.org.za/download/191.

Snowball, Jeanette, and Bruce Seaman. 2017. "A Guide to Using the South African Festival Economics Impact Calculator (SAFEIC)." South African Cultural Observatory. http://www.culture-developpement.asso.fr/wp-content/uploads/2017/09/safeicguide1.pdf.

Statistics South Africa. 2018. "Statistics by Place: Mafikeng." Statistics South Africa. Accessed 1 February 2017. http://www.statssa.gov.za/?page_id=993&id=mafikeng-municipality.

Toerin, Daan. 2020. "Tourism and Poverty in Rural South Africa: A Revisit." *South African Journal of Science* 16, nos. 1–2: 72–81.

Tyrrell, Timothy J., and Robert J. Johnston. 2001. "A Framework for Assessing Direct Economic Impacts of Tourist Events: Distinguishing Origins, Destinations and Causes of Expenditure." *Journal of Travel Research* 40, no. 1 (August): 94–100. https://doi.org/10.1177/004728750104000112.

Van Aalst, Irina, and Rianne Van Melik. 2012. "City Festivals and Urban Development: Does Place Matter?" *European Urban and Regional Studies* 19, no. 2 (April): 195–206. https://doi.org/10.1177/0969776411428746.

Van Zyl, Ciná. 2011. "A Model for Positioning Arts Festivals in South Africa." *South African Theatre Journal* 25, no. 3 (May): 181–96. https://doi.org/10.1080/10137548.2011.674681

Visser, Gustav. 2005. "Let's Be Festive: Exploratory Notes on Festival Tourism in South Africa." *Urban Forum* 16, no. 2 (April): 155–75.

4

Creative Yukon: Finding Data to Tell the Cultural Economy Story

Suzanne de la Barre

Introduction

The Arctic is a region facing rapid change due to globalization; the growth of governments and institutions; climate change (Southcott 2013); and an increasingly empowered Indigenous population that have negotiated land claims, fought for the right to self-government, and who are involved in reconciliation using diverse instruments, including community economic development by way of tourism (Hull, de la Barre, and Maher 2016). Arctic tourism is also impacted by changes brought about through the development of new tourism seasons (Rantala et al. 2019) and transformations occurring to the role of Arctic urban areas (Müller et al. 2020).

Similar to other places on the planet, the experience economy is having a significant impact on the development of the region's creative and cultural sectors (referred to in this chapter as the "cultural sector"), and has enriched placemaking and place-marketing processes. In the mid-1990s, the phrase "cultural economy" emerged in the social sciences and humanities and became a subject of scholarly investigation for two main reasons: first, as a result of interest in the culturalization of the economy; and second, to address the commodification and materialization of cultural consumption (Pratt 2008; Lash and Urry 1993). The cultural sector includes music, dance, visual arts, storytelling, ceremonies, rituals, and folklore, and provides a means for communities to enhance diverse place-based considerations (OECD 2014). These sectors also incorporate activities linked to hobbies, traditions, popular culture, art, and new media, and are recognized as drivers of economic growth that coincide with the rise of the creative class in urban areas (Florida 2004; Scott 1999, 2000, 2010).

There is a growing motivation to understand the way the creative sector engages social innovation, increases community resilience, generates positive social change and cross-cultural engagement, and affects economic diversification. Collins and Cunningham's (2017) recent volume provides a framework approach to understanding the cultural economy in the peripheral regions of the European Union and aims to stimulate future analysis and discussion. Focussing specifically on the Arctic and its specific peripheral features, Petrov's (2017, 2016, 2014, 2011, 2008, 2007; Petrov and Cavin 2013, 2017) assessment of the "other economies," including the cultural economy, provides a compelling story of how creative capital in its widest reading is likely to play a defining role in the regional transformation of remote areas. Among other things, he suggests that these sectors are greatly embedded in and dependent upon the internal capacity of communities (Petrov 2017). Others contribute evidence that the cultural economy offers a means for communities to leverage place-based concerns toward desired outcomes, including: (1) engaging the sector to strengthen regional and community resilience and to revitalize the economic and cultural life of remote and rural regions that suffer from economic dislocation and decline (Fleming 2009; Gabe 2007; Huggins and Clifton 2011; Leriche and Daviet 2010); and (2) supporting economic diversification objectives, for instance through entrepreneurship or tourism (Cloke 2007; Petrov 2007, 2008). Brouder (2012) presents a case in point in his study of northern Sweden, where tourism is deployed as a catalyst for innovative local development in "creative outposts" such as Jokkmokk.

The cultural economy connection to tourism research has been made by many, including Richards (2011), who points out that tourism is a significant force for economic growth in the field of culture and creativity. Commenting on the relationship between the creative and cultural industries and tourism in the Nordic context, the Nordic Council of Minsters (2018) claim that "although tourism is often not considered a creative and cultural industry, the industry is closely related to Nordic Arctic culture and the promotion of it. Tourism exposes visitors to Nordic Arctic culture, either through experiences or creative and cultural commodities or products, which in turn provide new sources of income" (17). Moreover, tourism benefits from resident-oriented cultural activities such as events and festivals, which often serve as the intermediary between culture and cultural tourism.

The World Tourism Organization (1985) defined cultural tourism as "movements of persons essentially for cultural motivations such as study tours, performing arts and cultural tours, travel to festivals and other events, visits to sites and monuments, travel to study nature, folklore or art, and pilgrimages" (6). However, Richards (2003) explains that defining cultural tourism is problematic owing to two factors: (1) challenges in defining "culture," including the multiple and diverse interpretations of the term cross-culturally, and (2) the different approaches used to define cultural tourism—for instance McKercher and Du Cros's (2002) motivational, experiential, and operational approaches. More recently, Richards (2011) deliberated on how the "creative turn" in tourism studies altered the way we understand the "drivers" of creativity in relation to both tourism producers and consumers.

In this chapter, creativity is investigated with the aim of identifying data that can also contribute to broader-based sector implications: for instance, cultural activities are also defined as "something to do" and as what brings "people together for reasons other than promoting the creative industries per se" (Mayes 2012, 7). Richards (2011) contributes to theoretical developments on the co-creative dynamics that exist in the tourism context. He points out that the increased commodification of everyday life is at stake when tourists are involved in a community's day-to-day cultural activities. He explains the way arts and creative activities are increasingly visible in the cultural tourism market, and that cultural tourism is a desirable market because it is generally high-spending tourism. It is a type of tourism that can also stimulate a destination's cultural activity, where local residents can also gain access to the benefits of cultural tourism activities and events. Richards underlines the growing link between cultural tourism and creativity ("creative tourism"), where the visitor engages in self-development and personal skills enhancement, and is involved in experiences with the local culture at the same time. These points are similarly discussed in chapter 10 of this volume, in which

Prince, Petridou, and Ioannides provide insight into the way artist clusters support co-placemaking and satisfy the needs of both residents and visitors—arguably an increasingly recognized necessity for tourism development in small places, perhaps even more so when these small places are also located in remote regions.

Referencing the challenges posed by commodification, Gibson (2012a, 8) asks: How can research on the cultural economy "be made social and not assume a capitalist-oriented language of firms, growth, employment and export, and instead value the communitarian purposes to which creativity can be put"? For Gibson, the dilemma persists when he examines a similar objective a few years later in his exploration of the role of academic intermediaries and their potential to advocate for progressive alternatives (2015). With these challenges, Gibson aims to intervene in neo-liberal, market-driven, and narrow economic development objectives that dominate typical approaches to outcomes, and he calls for more reflective examinations of the creative industries and their potential for transformative agendas. Smed Olsen et al. (2016) are among those who claim the creative industries contribute to more than just economic benefits: they promote personal development, educational objectives, and social inclusion. This work has implications for regional and community planning and policy-making based on correlation with creative capital, innovation, and economic growth, for instance. In a similar manner, and significant when it comes to defining features of the Arctic region, Indigenous leaders are voicing their support for an enhanced focus on the transformative values associated with economic development generally (Dolter 2017), and tourism specifically (Bunten 2010).

Nonetheless, the quest to better understand new development opportunities in the world's remote and sparsely populated regions are hindered by challenges associated with our lack of knowledge about the cultural sector. Petrov (2016) explains that "although instances of cultural economy in Arctic communities are easy to find, there is no systematic knowledge of its volume, characteristics and geography" (12). In light of the growing attention placed on new economies, including the creative economy, the desire for economic diversification, community (development) imperatives, and the existing knowledge gaps and data challenges, this chapter explores local data sources that might allow us to be more attentive to the cultural sector story in a peripheral place.

Canada's Yukon offers an apposite example of a sparsely populated, peripheral, northern, and polar place that is undergoing change and embracing new economic opportunities. At 482,223 square kilometres and with a population of 42,152 (Government of Yukon 2020), the territory is a vast place and home to relatively few people. It is a place whose settler history and economic activity have largely been driven by mining, as well as by Canada's national interests. Cries of "Gold! Gold! Gold!" and other echoes of the Klondike gold rush of 1898 still resonate across the territory. Using the Yukon as a case study, this chapter aims to explore the challenges related to measuring the cultural industry in peripheral areas and identifying local, embedded sources that can help us understand the dynamics and relationships it is implicated in. Specific questions that support this query include:

- What type of secondary source and "place-based data" is available and accessible that can help us tell the cultural economy story in peripheral places?

- What co-relationships can be employed to determine the broad symmetries involved in the development of the creative and cultural sectors and economic goals (e.g., tourism), as well as social objectives (e.g., well-being, resilience)?

The chapter also discusses the impact the cultural sector is having on the territory, including its contributions to residents' quality of life and community well-being, as well as the engagement between the cultural and tourism sectors.

The following section situates creativity in the periphery and features past research on the creative and cultural sector in relation to its expression in remote areas, and then specifically in the northern and circumpolar region. The chapter then provides an investigation of the issues under discussion through a case study of the Yukon Territory, situated in the northwestern-most corner of Canada. A section on methodology and a description of the data collected is then followed by a discussion based on the findings. The chapter concludes by proposing a research agenda going forward.

Where Is Creativity?

Scholars of creativity in peripheral regions, including Gibson (2012b), Collins and Cunningham (2017), and Petrov and Cavin (2017), all share the view that research on the creative economy has largely focussed on urban areas. Gibson's (2012b) research in Australia considered "what counts in small, remote, rural places—those places assumed by others to be 'uncreative' because of the histories of farming or manufacturing" (6). Collins and Cunningham (2017) uncover different facets and dimensions of the creative industries with the motivating rationale that they present crucial, underinvestigated opportunities for sustainable development in non-urban areas. Petrov and Cavin (2017) propose that the disregard for the way creativity functions in remote locations evidenced in the mainstream literature, which has focused almost exclusively on urban areas (e.g., Florida 2004, 2005) and mid-sized towns and cities (Margulies-Breibart 2013; Waitt and Gibson 2009), has led to the emergence of a literature on "creative peripheries."

Remote areas lie outside main centres of production and population, and are conceptualized in terms of the opportunities and challenges associated with the spatial arrangements that define them (Brown and Hall 2000). The literature on creative peripheries has questioned those perspectives that see proximity to urban areas as vital for the creative economy's development and success. Nonetheless, contributions that seek to better understand how creativity functions in the periphery describe a number of specific challenges linked to the context of remoteness. Gibson's (2012b) edited volume on research in Australia provides considerable insight into the features typically associated with the periphery. They are:

- Small size, so unable to tap into matters of critical mass and rate bases
- Ability to maintain visibility in larger markets
- Far-flung communities, lack of interconnectedness (among communities and to bigger places)

- Distance from key centres and scenes, and gatekeepers; both can also inhibit connections to international networks
- Relocation of talent to larger centres
- Dangers of parochialism
- Post-colonial setting

Specific obstacles may also include an aging population, low or transient social and cultural capital, limited infrastructure, economic marginalization, and constraints on information and governance (Brown and Hall 2000; Hall and Boyd 2005), all of which influence if and how the creative and cultural sectors are supported and promoted. Despite this inventory of obstacles, Petrov (2017) determines that, while the cultural economy in the periphery does follow a different template, it remains generally consistent with the basic principles of the creative capital theory. At the core of that theory is human capital and agency, which are both embedded in social networks and embraced by community (cf. Prince, Petridou, and Ioannides's chapter in this volume). Nonetheless, Petrov also suggests that human capital and agency may be more important for the transformation of peripheral areas than metropolitan ones.

Also contributing to the capital involved in deploying the potential of peripheral places are the place-values associated with them. Mayes (2012) pays tribute to the considerable value provided by remoteness, which can manifest as place distinctiveness and quirkiness. Similarly, others propose that remoteness can support the perception that people or products are "authentic"—uncorrupted, place-connected, and with a "disconnectedness from the machinations of urban capitalism" (Gibson, Luckman, and Willoughby-Smith 2012, 33). In this telling, the remote (frontier, colonial) context is upheld as a source of inspiration (30) and defined by a freedom from "city-based art fads" (Andersen 2012, 71). It is a space characterized by novelty, nostalgia, authenticity, and colonization-embedded and -empowered views of "untouched nature" (Cronon 1996; de la Barre 2013). Remote places also provide prospects for artists, entrepreneurs, and others to be "a big fish in a small pond," and as such they can offer increased chances for involvement in the governance and development aspects of creativity (e.g., policy, planning) (Verdich 2012, 138). Finally, as Gibson notes (2012a), in the tourism industry, "remoteness, marginality and difference can be brokered into a base for a distinctive and successful industry" (5).

There are other kinds of cultural questions involved in peripheral places, and these centre on the migration, interculturalism, and multi-culturalism of "new" settler groups who previously were not part of the discourse associated with the peripheral areas of the circumpolar North. If, as Richards (2011) claims, and resonating also with Carson et al. (2016), we are seeing the development of tourism as part of what challenges cur-rent representations of space, then the cultural industries play a crucial part in facilitating tourism's role in this spatial reconceptualization. In the Canadian context, the Arctic and its more general "northern tourism" context has typically been positioned as a nature and wilderness space. There are implications for how this northern space is (1) being reconfig-ured as a cultural space, and (2) how culture-based tourisms are (re)shap-ing that space, alongside Indigenization, reconciliation and post-colonial-ism (Hull, de la Barre, and Maher 2016).

Yukon's Creative and Cultural Economy

Yukon's communities are characterized by their differing degrees of re-moteness as well as their natural resource–based economies. Each informs these communities' determination as "path dependent" and also their eco-nomic diversification challenges, their limited or constrained human and economic resources, and their inadequate infrastructure and low popula-tions. Whitehorse has a population of 33,119, or about 75 per cent of the Yukon's overall population of 42,152 (Government of Yukon 2020). The city is within the shared traditional territory of the Ta'an Kwach'an Council, who signed their land claim and self-government agreements in 2002, and the Kwanlin Dün First Nation, who signed in 2005. This regional hub city is home to a number of art and cultural organizations, among them the Yukon First Nations Cultural and Tourism Association, the Yukon Art Society, the Yukon Artists at Work co-operative, the Northern Cultural Expressions Society, and the Association franco yukonnaise. There are also a growing number of not-for-profit associations representing the Yukon's diversity, such as the Canadian Filipino Association of the Yukon and the Chinese Canadian Cultural Association of Yukon . Whitehorse is also home to the Yukon Arts Centre, which hosts the only class "A" gallery space in northern Canada, a small business entrepreneurial and start-up community,[1] and various co-working spaces that have opened up within the last ten years.[2]

Dawson City is similarly positioned to tell a story about the emerging significance of the cultural and creative economies. The "city" is situated 532 kilometres north of Whitehorse, and has a population of 2,297 people (Government of Yukon 2020). As a national heritage site it is a major tourism centre that, up until recently, was known primarily for its role in the Klondike gold rush of 1898. Dawson City is increasingly understood for its contemporary positioning as a Yukon cultural "hub," and for the heritage and present-day significance of the Tr'ondëk Hwëch'in First Nation, who signed their land claim and self-government agreements in 1998. Arts and cultural organizations include the Dänoj Zho Cultural Centre, Yukon College's School of Visual Arts, and the Dawson City Arts Society, which is the managing body of the Klondike Institute of Art and Culture. Among its many festivals, the town is host to the Dawson City Music Festival, the International Short Film Festival, the Riverside Arts Festival, and the Shiver Winter Arts Festival. Dawson City's tourism industry has its biggest impact during the summer months, when the town's population booms with seasonal tourism workers and thousands of tourists, many of whom move through the town on their way to interior Alaska.

The Yukon provides a ripe context for this chapter as attention to its cultural sector has increased in the past two decades. An early sector assessment by Zanasi, Taggart, and Leaf (2004) led to the conclusion that the "cultural industries are one of the few bright spots in the Yukon economy. The sector is already an important part of the economy and it is suffused with optimism about its future" (ii). Extrapolating from national-level census statistics, they found that cultural employment had grown phenomenally in the years preceding their study and much faster than Canada as a whole—a 33 per cent increase from 1991 to 2001 for the national-level cultural labour force, compared to 100 per cent growth for Yukon. Growth was expected to continue. There has been no internal Yukon follow-up study to provide a contemporary contextualization for these assessments (a later study was initiated in 2016, but no final report was ever released); however, Petrov (2016) refers to a national study by Hills Strategies Inc. (2014), a private research firm that found that 4.62 per cent of the Yukon's 2011 labour force were in cultural occupations—the highest percentage in Canada.

Data Challenges in the Periphery

De Beukelaer (2014) has claimed that issues around data remain a major question for the cultural sector in general. He calls for "more locally grounded understanding of the significance of the creative economy—and culture in general" (94). Gibson (2012a) suggests that research on creativity in remote Australia requires new methods that enable us to understand the "geography of hidden, scattered creativity" (7). The case studies in his edited volume demonstrate both the complex challenges and opportunities of this sector, which are intrinsically related to remote geographical settings. Petrov (2016) considers the "fragmentary data and patchy knowledge" of the Arctic's "other economies" (2), and concludes that to understand them requires the use of diverse data-sets varying in scale, scope, and time coverage. He claims also that these economies tend to be more endogenous and embedded in peripheral locations, may have stronger internal linkages and multipliers, may generate more local development, and finally, because they have received only marginal and fragmentary attention, that there is limited data. In a later work, he also makes the case that standard methodologies used to analyze creative capital, a significant determinant for understanding how the creative sector functions, may not always be suited to non-metropolitan areas (Petrov 2017). Collins and Cunningham (2017) reiterate what others have determined before them with the claim that "one of the most challenging aspects of the understanding the creative economies and creative industries in peripheral regions is access to data that captures the extent, scope and unique characteristics of this sector" (4). Finally, and in a related vein, Carson et al. (2016), in their investigation of the opportunities outside traditional industries as a way for communities to become more innovative in addressing socio-economic decline, conclude that the lack of data may reinforce stereotypes and typical ways of dealing with different mobile populations (e.g., skilled workers, lifestyle migrants). It is against the backdrop of these data challenge proclamations that this study investigates identifying data sources and analyses that can help to tell the story of the cultural sector in a peripheral place.

Methodology and Findings

Case study research is an empirical inquiry of contemporary phenomena within a real life context (Yin 2014). It is a methodology that supports the in-depth exploration of one aspect of an issue or problem within a natural setting (Harland 2014), aims to answer "how" and "why" questions (Baxter and Jack 2008), and plays a significant role in advancing the knowledge base in a relevant field of study (Merriam 1998). Case study methodology has long been characterized as a weak approach among social science methods (Xiao and Smith 2006). However, early challenges to stereotypical perceptions of case study research state that it has been wrongly maligned, and they propose case studies as an effective way to refine general theory or effectively intervene in complex situations (Stoecker 1991). These types of assertions are found in tourism research (Xiao and Smith 2006), and they exist alongside claims that researchers can learn from previous case study research to support future research (Harland 2014). The present study and the approach used is a timely addition to the northern Canadian context as it also builds comparative opportunities with research in other remote locations, for instance in the Nordic countries (Power and Jansson 2008; Petridou 2011; Törnqvist 2011), and in Australia (Gibson 2012b).

The Yukon case study is concerned with gaining insight into what are the "locally embedded" secondary data sources available that will provide insight into the creative and cultural sector. To that end, data were collected and analyzed from four different sources:

1. Yukon economic, social, and cultural issues reports produced by diverse agencies

2. Government of Yukon–produced visitor guides

3. Government of Yukon–produced *Art Adventures on Yukon Time* guides

4. Government of Yukon arts support funds

YUKON STUDIES AND REPORTS

There are a number of reports about issues representing the Yukon's complex and interconnected economic, social, and cultural landscape produced by a variety of not-for-profits and governments (territorial and First Nations). Eleven reports were selected on the basis of being produced in the Yukon with the goal of examining a critical economic, social, or cultural issue or opportunity. The reports are summarily described in table 4.1. They are employed for their potential to inform what we know about the relationships that are influenced by and that are an influence on the cultural sector from perspectives that are embedded in place.

The perspectives are deployed so as to bring together views on the influence the cultural sector has on the economic, social, and cultural life of places. In light of the "transformational" benefits attributed to this sector, and given the existing literature that assigns different kinds of objectives and outcomes to the sector, particular attention was paid to the way the cultural sector was implicated as a way to foster positive change. The following correlation features were used:

- Support social inclusion—for instance connections to poverty reduction and intercultural learning strategies

- Contribute to First Nations cultural revitalization, healing, and wellness

- Northern uniqueness—for instance, that determines placemaking and as a way to create community, and for tourism related place-marketing

The economic correlation features aimed to also assess the valuation of transformational economic outcomes expected from economically related activity:

- Building community resilience

- Place-based (endogenous) employment and economic activity diversification

The reports highlight the expected instrumentalist rationales for supporting the creative industries, such as economic sector diversification,

Table 4.1. Yukon Reports

	OBJECTIVES	CORRELATION FEATURES	REPORT
		Cultural ways and being and relationship to wellness; cultural continuity, approaches, competence, and responsiveness	Forward Together: Yukon Mental Wellness Strategy – 2016–2026 (YG 2016)
		Franco-Yukon cultural tourism experiences and business development; celebration of Franco-Yukon culture; support cultural exchange and intercultural experience	Executive Summary: Feasibility Study from Homestay Vacations in the Yukon (AFY) (Chevalier 2013)
		Art and culture for creating inclusion and participating in society	A Better Yukon for All: Government of Yukon's Social Inclusion and Poverty Reduction Strategy (Westfall 2010; YG 2012)
		Cultural life as knowledge worker attraction requirement (e.g., arts, cultural vibrancy, and diversity)	Survey of Yukon's Knowledge Sector: Results and Recommendations (YRC) (Voswinkle 2012)
Creative and Cultural Sector	Transformational Economic Social Cultural	Develop and promote francophone arts and culture	Feasibility Study: Franco-Yukon Cultural Tourism Products RDÉE Yukon) (Binette 2011)
		Economic diversification through tourism and cultural industries; cultural revitalization (First Nations cultural centres), and arts infrastructure, Dawson City School of Visual Art (SOVA) as a means to support positive community change	Yukon Poverty Reduction Policies and Programs: Yukon (CCSD) (Edelson 2009)
		Use of culturally relevant arts, traditional crafts	Feasibility Study and Plan for Yukon First Nations Healing and Wellness Centre (YFNSGS) (Penny 2008)
		Economic development and diversification; First Nations Land Claims agreements and self-government objectives (e.g., heritage economic development); importance of and relevance of creative sector through tourism	Yukon's Cultural Labour Force (Zanazi, Taggart, and Leaf 2004)

increasing employability, and creating employment. However, there are also intersections with the cultural economy and industries in less obvious ways. Those intersections present instances that point to the way the cultural industries engage with social inclusion, community revitalizations (cultural, community), and increasingly play a role in the development and valuation of sectors other than those typically valued in natural resource–based contexts.

VISITOR GUIDES

The tourism sector's representation of culture as an asset is a way to assess the development and value of the creative and cultural sector. One way to examine this is to determine if there has been change from nature-based tourism to culture-based tourism. To do this, eleven issues of the Yukon tourism department–produced visitor guides were analyzed. Several high-frequency terms were identified and were themed along two main categories—culture and nature; these are itemized in table 4.2. While it is acknowledged that culture and nature can be viewed as a false dichotomy (Haila 2000; Selin 2003), and also that all tourism is cultural (Richards 2003), the dichotomized high-frequency terms are deployed in order to make useful distinctions for understanding the core amenity or asset promoted for the purposes of tourism.

The four earliest published Government of Yukon visitor guides were published inconsistently (1986, 1991, 1996, and 2001), and are only available in hard copy; they were accessed from government archives and were manually analyzed. Seven random issues produced between 2007 and 2018 were also analyzed; these are available electronically as PDFs, and the "find" feature was used to calculate frequency of word usage. Content difference was calculated on the basis of the number of times the high-frequency terms appeared in each of the visitor guides. Figure 4.1 presents the summary findings of the content analysis.

The high-frequency terms, while they fluctuate across time, consistently show an ever-increasing number of "culture" terms vis-à-vis the number of "nature" terms. However, it is the findings from the high-frequency terms analysis for 1986 (first guide available) and 2018 that tell the most worthwhile story. The data summary shows that the difference in culture and nature high-frequency terms in 1986 was 45.68 per cent; thirty-two

Table 4.2. Culture and Nature High-Frequency Terms

CULTURE	NATURE
Art(s) / Artist(s)	Adventure(s)
Creative	Nature / Natural
Culture(s) / Cultural	Outdoor(s)
Festival(s)	Wild
(Handi)craft(s)	Wilderness
Heritage	Wildlife

years later, the difference in culture and nature high-frequency terms is 17.48 per cent. While nature terms still dominate, there is a positive 28.2 per cent difference in how much emphasis is placed on "culture" in the visitor guides. As a potential indicator of change that has an influence on, or is influenced by, the cultural industries, these figures encourage us to ask if culture-related offerings are repositioning the tourist experience: Is the cultural sector transforming what has been a largely nature-based tourism destination into one that pays more attention to the presence and activities of humans—and by extension, those activities that give humans-in-place cause to reflect upon, make meaning of, enjoy, and express their world? In a similar vein, it may also suggest that the cultural sector is expanding visitor markets to the territory to include more culturally motivated visitors—if not exclusively or primarily motivated by culture, than at least as a significant complement to their nature-based motivations. An overarching question that arises from this analysis is whether or not changes that may be happening in relation to the type of tourism occurring in a destination point to other types of meaningful and non-tourism, community-embedded-related transformations?

ARTS FUNDING

Changes in the number of funding programs available to support artists and the development of the arts, the type of fund, and the amounts disbursed by the Arts Section of the Yukon government provides a means to assess aspects of the cultural sector. Figure 4.2 illustrates the changes in the number of funds, year created, and purpose of funding from 1983,

Figure 4.1. Visitor Guides 1986–2018—Summary

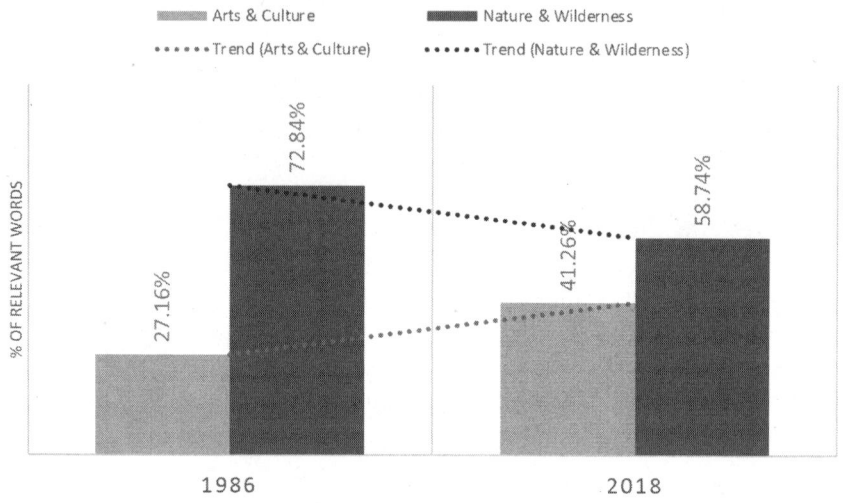

DIFFERENCE BETWEEN CULTURE AND NATURE (% OF RELEVANT WORDS)

Arts & Culture   Nature & Wilderness

Trend (Arts & Culture)   Trend (Nature & Wilderness)

% OF RELEVANT WORDS

27.16%   72.84%   41.26%   58.74%

1986   2018

Figure 4.2. Yukon Art Funds, 1983–2014

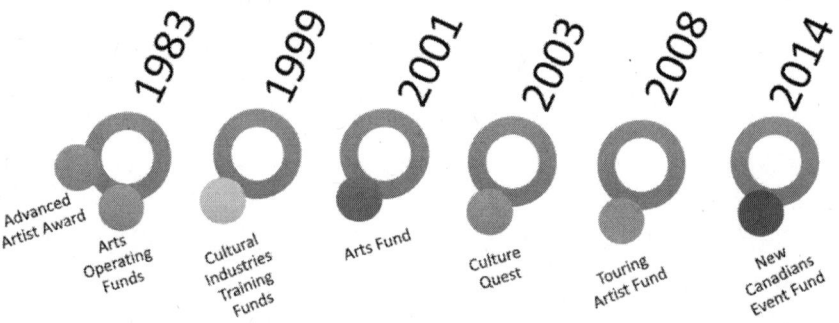

1983   1999   2001   2003   2008   2014

Advanced Artist Award   Arts Operating Funds   Cultural Industries Training Funds   Arts Fund   Culture Quest   Touring Artist Fund   New Canadians Event Fund

when the first two funds were created (Advanced Artist Award and Arts Operating Funds). One additional fund was created in each of the following years: 1999, 2001, 2003, 2008, and 2014. This brought the total of arts-related funds to seven.[3]

Figure 4.3 presents the estimated amount of funding available from all sources from 1983 to 2015 (Personal communication, Arts Section, Department of Tourism and Culture, Government of Yukon, 11 July 2016). Three distinct periods are apparent: (1) 1983 to 1988, with $20,000 disbursed using two funds; (2) 1988 to 1998, using the same two funds but with an increase of $40,000 per year, for a $60,000 annual disbursement; and (3) the period from 1999 to 2015, in which a new fund is created and added to the existing funds in 2001, 2003, 2008, and 2014, with a total of $1.65 million disbursed from the seven funds.

Art funds and the amounts presented are specifically from the Department of Tourism and Culture, and do not present the total number of funds or amounts of funding available to support the cultural sector from other sources. Other Government of Yukon funding examples include the Community Development Fund and the funding available through Lotteries Yukon. The latter disbursed $219,644 of its $318,676 Recreation Projects Program funding—or about 70 per cent—to cultural sector–related projects (*Yukon News*, 23 December 2016). The increase in Yukon government funding amounts, along with the type of funding available—including the objectives the fund is meant to support—will have had numerous impacts on the development of the arts and the cultural industries. Some insight into the number of artists and where to see and buy art can improve our understanding of these relationships.

Number of Artists and Places to See and Buy Art

The *Art Adventures on Yukon Time* publication results from a voluntary program and is not a comprehensive inventory of Yukon art makers and their art forms. Moreover, the program has changed over the years, and this impacts where art can be viewed or purchased. For instance, artists without studios were invited to participate in 2014; prior to that, only artists with studios could participate (Personal Communication, Arts Section, Department of Tourism and Culture, Government of Yukon, 11 July 2016). Figures 4.4 and 4.5 present summary results of the changes in

Figure 4.3. Yukon Art Funding Amounts Disbursed, 1983–2015

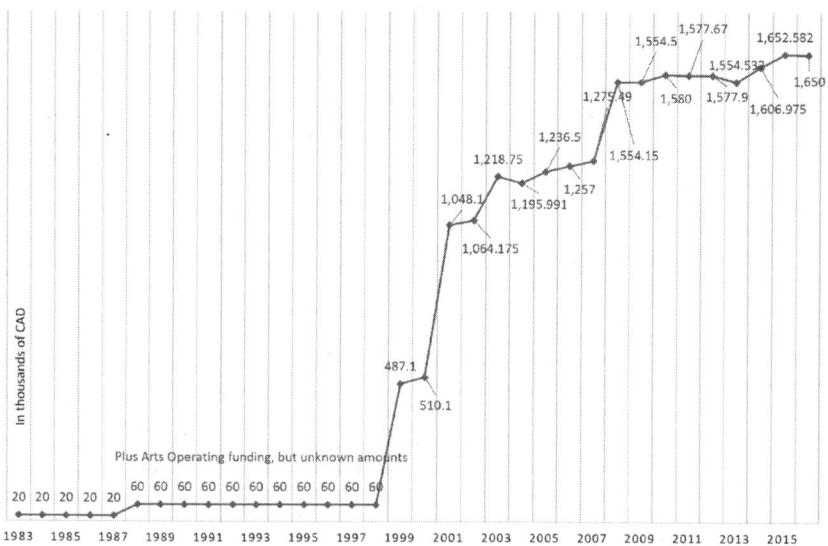

ART FUND AMOUNTS AWARDED ACROSS THE YEARS

the number of artists and where to see art calculated between 2000 and 2018 for three locations: Whitehorse, Dawson City, and all of the Yukon.

In the eighteen years since the guides have been produced, there are almost six times as many artists in Whitehorse, almost four times more artists in Dawson City, and almost five times as many artists in all of the Yukon.

For places to see and buy art, the increases are apparent at the Whitehorse and Dawson City levels, and especially significant from the "all Yukon" perspective. It should be noted that what is not factored into these preliminary presentations are the numerous other types of venues that have emerged in recent years, especially during the last decade, for the sale and display of arts and cultural productions: for example, craft fairs, farmers' markets (e.g., Fireweed Community Market), and "pop up" events (e.g., Etsy Made in Canada).

Figure 4.4. Number of Yukon Artists 2000–2018/19

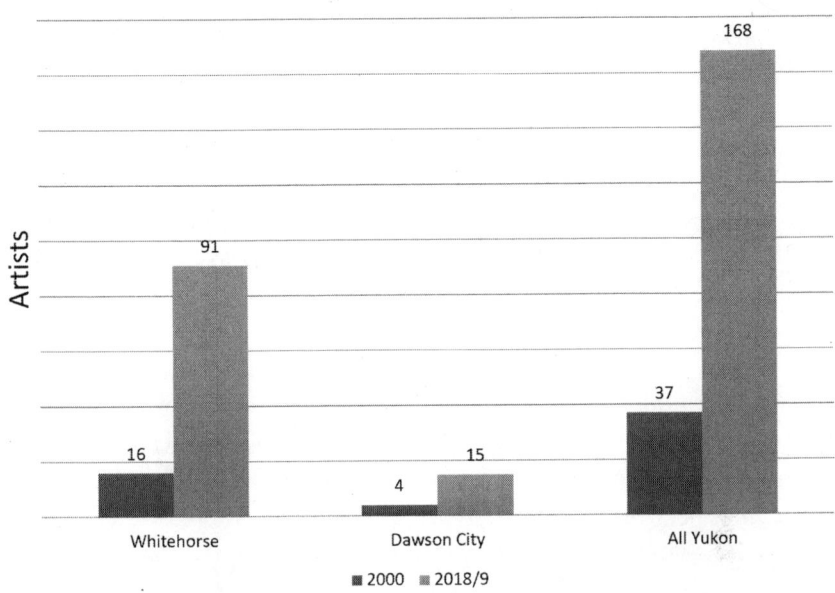

Figure 4.5. Where to see Art in Yukon 2000–2018/19

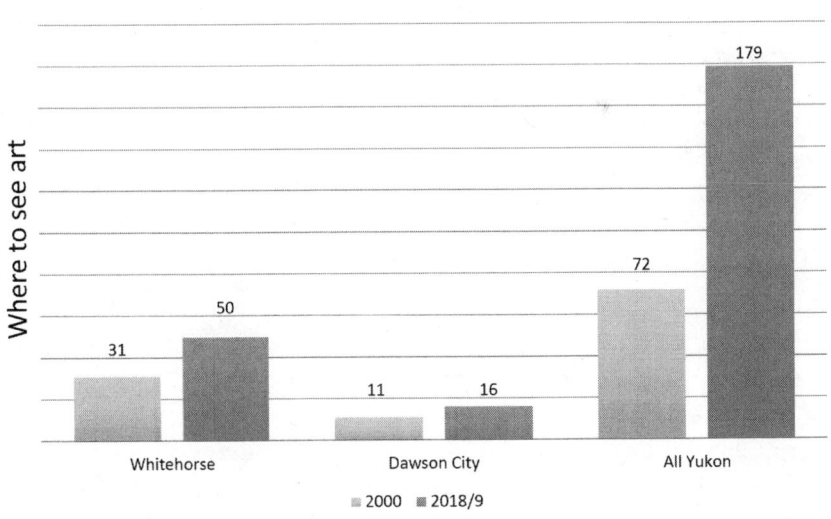

Discussion and Conclusion

There is a growing amount of research on creativity and the creative sector and its economies in the world's peripheral places, including the circumpolar region. Exploring new sources for place-embedded data can enhance our knowledge of the sector, how it functions, and what relationships it has. In addition to economic impacts, growth- or diversification-oriented information (e.g., tourism, lifestyle entrepreneurship), locally embedded data can also provide insight into the influence the creative sector has on broadly defined objectives of social and cultural enhancement, such as social inclusion and poverty reduction, and community development and revitalization.

A preliminary examination of select Yukon social and cultural reports demonstrates the potential to undertake assessment correlation analyses to better understand the interconnected nature of the arts-and-culture sector, and their support of diverse social, cultural, and economic development objectives. More rigorous approaches could provide a strategic rationale for supporting the cultural sector in order to advance transformational economic development objectives. They might also advance policy initiatives that support the cultural sector and its industries, while also having implications for meeting other social and cultural goals, such as social inclusion and community and cultural revitalization. Such policy formulation may also inform ways of achieving what many governments present as strategic governance goals that depend on the ability of government departments to issue cross-department goals and objectives. In this sense, pulling reports out of boxes and dusting them off with the intention of analyzing them for the way social, cultural and economic issues and their solutions are connected to the cultural sector could provide significant insights for developing multi-purpose policy, for achieving better inter-departmental collaboration, and for meeting overarching goals of community and economic development.

While the nature-culture divide may well be a social construction, the reference points are typically employed in tourism promotional materials in the peripheral areas of the circumpolar world. Nature-based tourism has dominated the circumpolar region internationally, and it is arguably only in the last decade, and specific to Indigenous empowerment movements and growing curiosity about original peoples, that cultural tourism has been given much attention—even if the implications are much wider than just the significant Indigenous cultural reference points. Given the connection between the cultural sector and tourism, identifying changes in the nature and type of tourism in peripheral regions can also add to our understanding of more than just "emerging" or "new" economies. These changes can also inform our understanding of spatial arrangements and allow us to reflect upon meaningful demographic and intercultural dynamics. Arguably, these changes also encourage a revised view of the North: for instance, as an empowered Indigenous space, and one that also has the potential to narrate the circumpolar region as more than an assumed bicultural space. The latter notion has up to now been primarily defined by frontier, colonial, and settler relations that persist due to a perceived passive population (as opposed to a population that was systemically oppressed). Indeed, identifying these tourism-related changes encourages our ability to embrace cultural constructions that are complex, multi-cultural, and diverse. They also make possible East-West relationships across the circumpolar region to interrogate the historical legitimacy and colonization-embedded rationale for South-to-North dependencies.

Increased funding over time suggest there are significant consequences to the development of the cultural sector, including acknowledging the growing significance of cultural producers and new opportunities for consumers. The relative growth in the number of artists and places to see and buy art appears to correspond almost directly with increases in arts-and-culture funding across time. Similar findings were documented by Zanasi, Taggart, and Leaf (2004), who already in 2004 pointed out that the Yukon's spending on arts and culture led to the "phenomenal growth of the cultural sector" (22). They also lamented the lack of "hard numbers," which, they argued, interfered with the effort to establish whether or not cultural organizations were experiencing decreases in specific cultural funding. A partial assessment of funding for arts and culture is not an effective way to make conclusions regarding relationships or policy recommendations; other funds, such as Yukon Lotteries, would suggest that a more complete picture of funding for arts and culture would prove useful.

These four analyses offer examples of locally embedded data sources that help us to tell the cultural sector story. They enhance insights into several relationships, including 1) the cultural sector and its ability to contribute to social, cultural, and diverse types of economic goals; 2) the connection between cultural content represented by tourism place-marketing in relation to nature content, and the possibility that these relationships may help us also understand Indigenous cultural and other empowerment (e.g., political, economic), and meaningful transformations to the increasing multicultural diversity of the territory; and 3) the relationship between funding support and cultural capital, represented in the number of arts-and-culture producers and consumers' opportunities to access their productions. If, as Petrov (2016) declares, our future task is to improve our limited understanding of the cultural sector, its economies and its relationships with other sectors, community well-being, and sustainable development, then efforts to broaden where we look for and find data—and how we assess it—play a vital role in support of our ability to achieve these goals.

NOTES

1   See https://www.startupcan.ca/2019/05/2019-startup-canada-award-winners-announced-in-whitehorse/.

2   See http://yukonstruct.com/.

3   More information on these funds is available at https://yukon.ca/en/arts-and-culture.

References

Andersen, Lisa. 2012. "Magic, Light, Silver City: The Business of Culture in Broken Hill." In *Creativity in Peripheral Places: Redefining the Creative Industries*, edited by Chris Gibson, 71–86. New York: Routledge.

Baxter, Pamela, and Susan Jack. 2008. "Qualitative Case Study Methodology: Study Design and Implementation for Novice Researchers." *Qualitative Report* 13, no. 4: 544–59.

Binette, Sylvie, Christiane Boisjoly Inc., and Luigi Zanasi. 2011. "Feasibility Study: Franco-Yukon Cultural Tourism Products." http://www.afy.yk.ca/secteurs/documents/en/d314-final-report.pdf.

Brouder, Patrick. 2012. "Creative Outposts: Tourism's Place in Rural Innovation." *Tourism Planning and Development* 9, no. 4: 383–96.

Brown, Frances, and Derek Hall. 2000. *Tourism in Peripheral Areas: Case Studies.* Clevendon, UK: Channel View.

Bunten, Alexis. 2010. "More Like Ourselves: Indigenous Capitalism through Tourism." *American Indian Quarterly* 34, no. 3: 285–311.

Carson, Doris, Jen Cleary, Suzanne de la Barre, Marco Eimermann, and Roger Marjavaara. 2016. "New Mobilities—New Economies? Temporary Populations and Local Innovation Capacity in Sparsely Populated Areas." In *Settlements at the Edge: Remote Human Settlements in Developed Nations*, edited by Andrew Taylor, Dean Carson, Prescott Ensign, Lee Huskey, and Rasmus Ole Rasmussen, 178–206. Cheltenham, UK: Edward Elgar Publishing.

Chevailier, Stephanie. 2013. "Executive Summary: Feasibility Summary Homestay Vacations in the Yukon." https://www.afy.yk.ca/secteurs/documents/en/d528-executive-summary-feasibility-study-homestay-vacations-in-the-yukon.pdf.

Cloke, Paul. 2007. "Creativity and Tourism in Rural Environments." In *Tourism Creativity and Development*, edited by Greg Richards and Julie Wilson, 37–47. London: Routledge.

Collins, Patrick, and James A. Cunningham. 2017. *Creative Economies in Peripheral Regions.* Cham, CH: Springer International.

Cronon, William, ed. 1995. *Uncommon Ground: Toward Reinventing Nature.* New York: W. W. Norton and Company.

De Beukelaer, Christiaan. 2014. "The UNESCO/UNDP 2013 Creative Economy Report: Perks and Perils of an Evolving Agenda." *Journal of Arts Management, Law, and Society* 44, no. 2: 90–100.

de la Barre, Suzanne. 2013. "Wilderness and Cultural Tour Guides, Place Identity, and Sustainable Tourism in Remote Areas." *Journal of Sustainable Tourism* 21, no. 6: 825–44.

Dray, Chris. 2004. "Cultural Spaces and Places: Cultural Tourism Infrastructure in Downtown Whitehorse." Yukon Arts Centre Corporation, Roundtable on the Cultural Economy, Artspace North, March 2014.

Edelson, Natalie. 2009. *Poverty Reduction Policies and Programs: Yukon.* Social Development Report Series. Whitehorse: Canadian Council on Social Development.

Fleming, Rachel C. 2009. "Creative Economic Development, Sustainability and Exclusion in Rural Areas." *Geographical Review* 99, no. 1: 61–80.

Florida, Richard. 2004. *The Rise of the Creative Class and How It's Transforming Work, Leisure, Community and Everyday Life.* New York: Basic Books.

———. 2005. *Cities and the Creative Class.* New York: Routledge.

Gabe, Todd. 2007. "Foreword: Special Issue on Opportunities and Challenges Facing the Rural Creative Economy." *Agricultural and Resources Economics Review* 36, no. 1: iii–iv.

Gibson, Chris. 2015. "Negotiating Regional Creative Economies: Academics as Expert Intermediaries Advocating Progressive Alternatives." *Regional Studies* 49, no. 3: 476–9.

———. 2012a. "Cultural Economy: Achievements, Divergences, Future Prospects." *Geographical Research* 50, no. 3 (August): 282–90.

———. 2012b. *Creativity in Peripheral Places: Redefining the Creative Industries.* New York: Routledge.

Gibson, Chris, Susan Luckman, and Julie Willoughby-Smith. 2012. "Creativity without Borders? Rethinking Remoteness and Proximity." In *Creativity in Peripheral Places: Redefining the Creative Industries,* edited by Chris Gibson, 25–38. New York: Routledge.

Government of Yukon. 2012. *A Better Yukon for All: Government of Yukon's Social Inclusion and Poverty Reduction Strategy.* Whitehorse: Government of Yukon, Department of Health and Social Services.

———. 2013. Visual Arts and Craft Steering Committee. *Yukon Visual Arts and Craft Strategy.* Whitehorse: Department of Economic Development.

———. 2016. *Forward Together: Yukon Mental Wellness Strategy, 2016–2026.* Whitehorse: Department of Health and Social Services. http://www.hss.gov.yk.ca/pdf/mentalwellnessstrategy.pdf.

———. 2020. *Population Report First Quarter, 2020* (March). Whitehorse: Yukon Bureau of Statistics. https://yukon.ca/sites/yukon.ca/files/ybs/populationq1_2020.pdf.

Haila, Yrjö. 2000. "Beyond the Nature-Culture Dualism." *Biology and Philosophy* 15, no. 2: 155–75

Hall, Colin M., and Stephen W. Boyd. 2005. *Nature-Based Tourism in Peripheral Areas: Development or Disaster?* Clevendon, UK: Channel View.

Harland, Tony. 2014. "Learning about Case Study Methodology to Research Higher Education." *Higher Education Research & Development* 33, no. 6: 1113–22.

Hills Strategies Inc. 2014. "Artists and Cultural Workers in Canada's Provinces and Territories," *Statistical Insights on the Arts* 12, no. 3. http://hillstrategies.com/resource/artists-and-cultural-workers-in-canadas-provinces-and-territories/.

Dolter, Brett. 2017. "Sustainable Inclusive Development: Carol Anne Hilton's Concept of Indigenomics Is the Next Transformative Step in Economics." *Alternatives Journal* 43, no. 1: 56.

Huggins, Robert, and Nick Clifton. 2011. "Competitiveness, Creativity, and Place-Based Development." *Environment and Planning A: Economy and Space* 43, no. 6: 1341–62. https://doi.org/10.1068%2Fa43559.

Hull, John, Suzanne de la Barre, and Pat Maher. 2016. "Peripheral Geographies of Creativity: The Case for Aboriginal Tourism in Canada's Yukon Territory." In *Tourism and Indigeneity in the Arctic,* edited by Arvid Viken and Dieter K. Müller, 157–81. Bristol, UK: Chanel View.

Lash, Scott, and John Urry. 1993. *Economies of Signs and Space.* London: Sage.

Leriche, Frédéric, and Sylvie Daviet. 2010. "Cultural Economy: An Opportunity to Boost Employment and Regional Development?" *Regional Studies* 44, no. 7 (August): 807–11.

Margulies-Breitbart, Myrna, Mark Boyle, Donald Mitchell, and David Pinder. 2013. *Creative Economies in Post-Industrial Cities: Manufacturing a (Different) Scene.* Farnham, UK: Ashgate.

Mayes, Robin. 2012. "Postcards from Somewhere: 'Marginal' Cultural Production, Creativity, and Community." In *Creativity in Peripheral Places: Redefining the Creative Industries,* edited by Chris Gibson, 11–24. New York: Routledge.

McKercher, Bob, and Hilary Du Cros. 2002. *Cultural Tourism: The Partnership between Tourism and Cultural Heritage Management.* New York: Haworth Press.

Merriam, Sharan. 1998. *Qualitative Research and Case Study Applications in Education.* San Francisco: Jossey-Bass Publishers.

Müller, Dieter K., Doris A. Carson, Suzanne de la Barre, Brynhild Granås, Gunnar Thór Jóhannesson, Gyrid Øyen, Outi Rantala, Jarkko Saarinen, Tarja Salmela, Kaarina Tervo-Kankare, and Johannes Welling. 2020. *Arctic Tourism in Times of Change: Dimensions of Urban Tourism.* TemaNord 2020: 529. Copenhagen: Nordic Council of Ministers. https://pub.norden.org/temanord2020-529/.

Nordic Council of Ministers. 2018. *Arctic Business Analysis: Creative and Cultural Industries.* Copenhagen: Nordic Council. http://norden.diva-portal.org/smash/get/ diva2:1175681/FULLTEXT01.pdf

OECD. 2014. *Tourism and the Creative Economy.* Paris: OECD Publishing. http://dx.doi. org/10.1787/9789264207875-en.

Penny, Meyers Norris. 2008. *Feasibility Study and Plan for Yukon First Nations Healing and Wellness Centre.* Prepared for the Yukon First Nations Self Government Secretariat.

Petridou, Evangelia. 2011. *Conducting Creativity in Central Sweden: A Path towards Territorial Cohesion.* Paper presented at Innovation Processes and Destination Development in Tourist Resorts, a conference organized by the Regional Studies Association Research and Network on Tourism and Regional Development. Östersund, Sweden, 30 March–1 April.

Petrov, Andrey. 2007. "A Look Beyond Metropolis: Exploring Creative Class in the Canadian Periphery." *Canadian Journal of Regional Science* 30, no. 3: 359–86.

———. 2008. "Talent in the Cold? Creative Capital and the Economic Future of the Canadian North." *Arctic* 61, no. 2 (June): 162–78.

———. 2011. "Beyond Spillovers: Interrogating Innovation and Creativity in the Peripheries." In *Beyond Territory: Dynamic Geographies of Innovation and Knowledge Creation,* edited by Harald Bathelt, Maryann Feldman, and Dieter Kogler, 168–90. New York: Routledge.

———. 2014. "Creative Arctic: Towards Measuring Arctic's Creative Capital." In *Arctic Yearbook 2014*, edited by Lassi Heininen, Heather Exner-Pirot, and Joel Plouffe, 149–66. Akureyri, IS: Northern Research Forum.

———. 2016. "Exploring the Arctic's 'Other Economies': Knowledge, Creativity and the New Frontier." *Polar Journal* 6, no. 1: 51–68.

———. 2017. "Human Capital and Sustainable Development in the Arctic: Towards Intellectual and Empirical Framing." In *Northern Sustainabilities: Understanding and Addressing Change in the Circumpolar World*, edited by Gail Fondahl and Gary Wilson, 203–20. Cham, CH: Springer Polar Sciences.

Petrov, Andrey, and Philip Cavin. 2013. "Creative Alaska: Creative Capital and Economic Development Opportunities in Alaska." *Polar Record* 49, no. 4: 348–61.

———. 2017. "Creating New Path Creative Capital: Theories and Evidence in the Northern Periphery." *Journal of Rural and Community Development* 12, nos. 2–3: 127–42.

Power, Dominic, and Johan Jansson. 2008. "Outside In: Peripheral Cultural Industries and Global Markets." In *Mobility and Place: Enacting Northern European Peripheries*, edited by Jørgen Ole Bærenholdt and Gunnar Thor Jóhannesson, 167–77. Farnham, UK: Ashgate.

Pratt, Andy. 2008. "Creative Cities: the Cultural Industries and the Creative Class." *Geografiska Annaler: Series B, Human Geography* 90, no. 2: 107–17.

Rantala, Outi, Suzanne de la Barre, Brynhild Granås, Gunnar Þór Jóhannesson, Dieter K. Müller, Jarkko Saarinen, Kaarina Tervo-Kankare, Patrick T. Maher, and Maaria Niskala. 2019. *Arctic Tourism in Times of Change: Seasonality*. TemaNord 2019: 528. Copenhagen: Nordic Council of Ministers. https://www.norden.org/en/publication/arctic-tourism-times-change-seasonality-0.

Richards, Greg. 2003. "What Is Cultural Tourism?" In *Erfgoed voor Toerisme*, edited by Ad van Maaren. Nationaal Contact Monumenten. Retreived from https://www.academia.edu/1869136/What_is_Cultural_Tourism.

———. 2011. "Creativity and Tourism: The State of the Art." *Annals of Tourism Research* 38, no. 4: 1225–53.

Scott, Allen. 1999. "The Cultural Economy: Geography and the Creative Field." *Media, Culture and Society* 21, no. 6: 807–17.

———. 2000. *The Cultural Economy of Cities*. London: Sage.

———. 2010. "Cultural Economy and the Creative Field of the City." *Geografiska Annaler: Series B, Human Geography* 92, no. 2: 115–30.

Selin, Helaine, ed. 2003. *Nature across Cultures: Views of Nature and the Environment in Non-Western Cultures*. Boston: Kluwer Academic.

Smed Olsen, Lise, Anna Berlina, Luneisja Jungsberg, Nelli Mikkola, Johanna Roto, Rasmus Ole Rasmussen, and Anna Karlsdottir for the Nordic Centre for Spatial Development. 2016. "Sustainable Business Development in the Nordic Arctic." Stockholm: NORDREGIO. http://www.nordregio.org/publications/sustainable-business-development-in-the-nordic-arctic/.

Southcott, Chris. 2013. "Globalization and Rural Change in Canada's Territorial North."
In *Social Transformation in Rural Canada: New Insights into Community, Cultures, and Collective Action*, edited by John R Parkins and Maureen Reed, 43–66. Vancouver: UBC Press.

Stoecker, Randy. 1991. "Evaluating and Rethinking the Case Study." *Sociological Review* 39, no. 1: 88–112.

Törnqvist, Gunnar. 2011. *The Geography of Creativity*. Cheltenham, UK: Edward Elgar Publishing.

Verdich, Madeleine. 2012. "Creative Migration? The Attraction and Retention of the 'Creative Class' in Launceston, Tasmania." In *Creativity in Peripheral Places: Redefining the Creative Industries*, edited by Chris Gibson, 129–40. New York: Routledge.

Voswinkel, Stefan. 2012. *Survey of Yukon's Knowledge Sector: Results and Recommendations*. Whitehorse: Yukon Research Centre, Yukon College.

Waitt, Gordon, and Chris Gibson. 2009. "Creative Small Cities: Rethinking the Creative Economy in Place." *Urban Studies* 46, nos. 5–6: 1223–46.

Westfall, Rachel. 2010. *Dimensions of Social Inclusion and Exclusion in Yukon, 2010*. Whitehorse: Department of Health and Social Services. http://www.deslibris.ca/ID/226436.

World Tourism Organization. 1985. *The State's Role in Protecting and Promoting Culture as a Factor of Tourism Development*. Madrid: WTO.

Xiao, Honggen, and Stephen L. J. Smith. 2006. "Case Studies in Tourism Research: A State-of-the-Art Analysis." *Tourism Management* 27, no. 5: 738–49.

Yin, Robert. 2014. *Case Study Research: Design and Methods*, 3rd ed. Thousand Oaks, CA: Sage.

Yukon News. 2016. Lotteries Yukon advertisement (p. 26), 23 December.

Zanasi, Luigi, Cathy Busby, Sherri Irvin, Malcolm Taggart, and Yukon Arts Centre. 2000. *Developing the Arts Scene: Art Education as a Cultural Tourism Industry*. https://yukonomics.ca/reports/Arts_Education.PDF.

Zanasi, Luigi, Malcolm Taggart, and Wanda Leaf. 2004. *The Yukon's Cultural Labour Force*. Whitehorse: Yukon Arts Centre Corporation. http://www.yukonomics.ca/reports/Yukon_Cultural_Labour_Force.pdf.

When Our Ship Comes In: The Cultural Impact of Cruise Tourism on Northern Canadian Communities

M. Sharon Jeannotte

Introduction

In August 2016, the *Crystal Serenity*, a thousand-passenger luxury cruise ship, sailed through the Northwest Passage in Canada's North. Although smaller expedition cruise ships have been plying these waters for a number of years, the *Crystal Serenity* is the largest passenger ship to make the passage, raising the stakes for a region that is already trying to cope with the impacts of climate change due to rapidly melting sea ice. Its passage was widely covered in the Canadian and international media (Landriault 2019), and it has spurred governments at all levels to consider possible catastrophic occurrences—such as oil spills and other marine accidents—in the remote and sensitive Canadian Arctic. But beyond concerns such as the lack of infrastructure to cope with large vessels and the difficulties of monitoring and regulating marine traffic in such a remote area, there lie many less tangible issues that pose long-term challenges for governance and quality of life in Canada's northern regions. These include the potential social, economic, and environmental disruptions caused by huge incursions of tourists, as well as the cultural effects that mass tourism could have on small, remote Indigenous communities in the Canadian Arctic. While these impacts can be addressed to a certain extent by policy interventions, the complex history of the region and the unique world views of the peoples who inhabit it influence not only the short-term costs and benefits of cruise tourism, but also the longer-term sustainability of Arctic cultures and ecosystems.

This chapter explores risks to environmental, economic, social, and cultural sustainability in these communities, as well as the complicated relationship between creative practices, cultural tourism, Indigenous/ Inuit values and world views, and local planning practices in this part of Canada. It will also discuss some of the ethical and practical implications of Inuit–cruise passenger interactions, and explore the adaptive capacities of these communities to cope with larger and more frequent cruise tourism incursions.

Culture, Creativity, Sustainability, and Tourism—A Complicated Relationship

The *Crystal Serenity*'s passage highlights the complicated relationship between culture, creativity, sustainability, and tourism, both in general terms and in the specific context of the North—topics that have also been discussed by de la Barre and by Aquino and Burns in other chapters in this volume. Definitions of each of these elements are contested, and when they are combined, they create a definitional and analytical maze. What is "sustainable tourism"? What is "culture" and how does it fit within the contested field of "sustainable tourism"? What is "creative tourism"? And how does creative tourism fit within the unique environment of Canada's North, where climate change threatens to erode the basis of the region's culture and the creative practices embedded in this fragile place?

The concept of "sustainable tourism" contains a number of inherent contradictions and paradoxes. Sustainable tourism began as a form of "alternative tourism," which was put forward as an alternative to conventional mass tourism in the 1980s (Weaver 2015). However, several contradictions have emerged within the concept and practice of alternative tourism. Many of the expectations for such local, small-scale, community-based tourism have failed to materialize. Often, this type of tourism is dominated by elites and fails to meet the economic needs of host communities because of its small scale. It also frequently relies upon the facilities and services of mass tourism systems, especially in the area of transportation. Indeed, it can become an appendage of mass tourism, rather than a true alternative (Weaver 2015, 13–14).

The United Nations World Tourism Organization (UNWTO 2014) has defined sustainable tourism as "tourism that takes full account of its current and future economic, social, and environmental impacts, addressing the needs of visitors, the industry, the environment, and host communities." It suggests that besides making optimal use of environmental resources, sustainable tourism should also "respect the socio-cultural authenticity of host communities" and "ensure viable, long-term economic operations" (UNWTO 2014). However, as Butler (2015) has argued, definitions such as these contain within them a series of paradoxes. As with "alternative tourism," a sustainable tourism site or destination may be too specialized to meet expectations for economic development or may be degraded or damaged by too many visitors. But the paradox most relevant to the topic of cruise tourism in Canada's North is that "most tourism is, and will remain unsustainable because it involves travel" and that all forms of travel "beyond walking, sailing, cycling, or horse drawn forms involve the use of non-sustainable energy sources" (Butler 2015, 74). Mass cruising in Canada's North relies on motorized vessels burning large amounts of fossil fuels that contribute not only to greenhouse gas emissions and melting sea ice, but also, potentially, to environmental damage from fuel leaks resulting from marine accidents.

Culture is recognized as an important element of the sustainable tourism mix. As UNESCO has stated, "Cultural and natural heritage sites, intangible cultural heritage, performing arts and museums are among the many interests that engage tourists and thereby generate revenues, employment, investments, and social benefits" (UNESCO 2017). However, culture can mean different things to different people, and it is always necessary to state what definition of "culture" one is using when discussing how it fits within the idea and practice of sustainable tourism. Smith (2016) suggests that Raymond Williams's notion of culture "being about a whole way of life as well as the arts and learning" is a useful place to start; she also argues that the definition of "culture" used in tourism marketing has expanded from a narrow focus on the arts and heritage to encompass such everyday activities as "shopping, football, events, and all forms of food and drink" (Smith 2016, 1). In places with large Indigenous populations, she adds that "few indigenous and tribal peoples of the world sit around discussing the meaning of culture. Instead, they live and breathe it" (2).

In light of concerns about sustainable tourism, it is worth reviewing the various understandings of culture in fragile environments. Based on their research on culture and sustainability, Duxbury and Jeannotte (2013) identified five separate ways in which culture is treated: (1) as capital; (2) as process and way of life; (3) as a vehicle for sustainable values; (4) as creative expression; and (5) as complex networks of interdependent social and economic systems. In the case of cruise tourism in Canada's North, all five categories of culture are relevant. Canada's North has a stock of both tangible and intangible cultural capital that it has inherited and wishes to preserve, and the way of life of its inhabitants is very much a "formalization of practices by individuals and/or communities as they adjust to, survive and prosper in special contexts" (Rana and Piracha 2007, 22). The philosophy, ethics, Traditional Knowledge, and values of Indigenous Peoples in Canada's North have helped sustain their communities in the past and continue to guide them in their way of life. Creative expression, in the form of the visual and performing arts, is part of the fabric of northern communities, as well as being a major source of revenue (Milne, Ward, and Wenzel 1995; Nordicity Group and Uqsiq Communications 2010; Big River Analytics 2017). Finally, the complex relationships and feedback loops between culture and socio-economic systems in the North have created systems of adaptive renewal that contribute to community resilience (Lemelin et al. 2012; Bennett and Lemelin 2015).

Creative tourism, as defined by UNESCO, is "travel directed toward an engaged and authentic experience, with participative learning in the arts, heritage, or special character of a place, and it provides a connection with those who reside in this place and create this living culture" (UNESCO 2006, 3, quoted in Richards 2011, 1237). Richards (2011) argues that the embeddedness of creative knowledge and skills is one of the main attractions of creative tourism, which packages cultural capital—both tangible and intangible—into an "experience" that is often a key part of the local symbolic economy (1228). Further, as tourism becomes a more important element of the local economy, it can begin to reshape the traditional culture by encouraging new creative dimensions and forms of expression (1236). Tourists in this context become change agents, as "consumers can influence the politics of the product and production process through . . . their consumption behaviour" (Verbeek and Mommaas 2008, 639).

A broad overview of the history, climate, geography, socio-economic, and political status of Canada's North will set the stage for how these elements come together in the specific context of cruise tourism.

Background on Canada's North

Canada's three northern territories—Nunavut, the Northwest Territories, and Yukon—together comprise almost 4 million square kilometres or approximately 40 per cent of Canada's total area. Politically, they are self-governing but have less autonomy than the Canadian provinces and rely heavily upon the federal government for infrastructure and services. The temperature in Iqaluit, the capital of Nunavut, the territory that has experienced the most cruise tourism, averages -27 degrees Celsius in January and rises to between 2 and 8 degrees Celsius between June and September (Nunavut Tourism, n.d.a). However, since 1980, air temperatures over the Arctic Ocean have increased by 1 to 2 degrees Celsius due to climate change, and the length of the Arctic summer has increased by five to ten weeks throughout the Arctic (Hoag 2016).

The total population of the three territories in 2016 was only 113,604, with about 50 per cent of the inhabitants claiming Aboriginal ancestry (Statistics Canada, n.d.).[1] This population lives in small communities stretched across an expanse greater than the distance between Dublin and Moscow. The percentage of the population aged fourteen and under is almost twice as high in Nunavut as in the rest of Canada (Statistics Canada, n.d.; Southcott and Walker 2015, 28).

Prior to contact with Europeans, the social economy of the North was based on hunting and gathering and was heavily dependent on environmental conditions, which dictated the movements of animal species such as bears, caribou, and whales, as well as the relative abundance or scarcity of the animal harvest. As Southcott and Walker note, "the fur trade introduced a new system of relations to the region that can best be called pre-industrial colonialism" (2015, 22). This system, along with the whaling industry, which began to develop in the eastern Arctic in the eighteenth century, did not eradicate the traditional economy but transformed it from a purely subsistence basis to one that "combined subsistence with a dependence on servicing the economic needs of primarily European populations" (24).

These relations continued under the subsequent system of industrial colonialism, based on mining and petroleum, which started with the Klondike gold rush in the Yukon in the late nineteenth century and the discovery of large oil and gas deposits in the Mackenzie Delta region of the Northwest Territories in the 1960s (Southcott and Walker 2015, 24–5). In both cases, these industries relied upon workers imported from the South and excluded Indigenous Peoples, who continued to rely upon their traditional subsistence economy. In the 1950s, the federal government began to move Inuit populations into permanent settlements (sometimes by coercive means) to facilitate the delivery of health, educational, and social services. In the new settlements, the colonial pattern continued in the form of externally controlled administration (Haalboom and Natcher 2013, 363) and importation of workers from the South to deliver services (Southcott and Walker 2015, 26). Along with high unemployment rates among the Aboriginal population of the northern territories, other legacies of colonialism in the North include high suicide rates and relatively lower levels of educational achievement (Canadian Northern Economic Development Agency 2016; Conference Board of Canada 2014; Hicks 2015).

The current economy in Canada's North is "mixed," with the subsistence activities of the original hunting-and-gathering societies coexisting with private-sector resource extraction and the activities of governments—health, education, policing, and social services (Southcott and Walker 2015). A significant social economy, based on co-operatives or other collectively owned and operated organizations, has arisen in the North as a reflection of the region's traditional values of sharing and communal activity (Southcott and Walker 2015; MacPherson 2015). A 2008 census counted almost twelve hundred social economy organizations in the three territories, with the highest numbers operating in the areas of tourism, sport and recreation, social services, and arts and culture (Southcott and Walker 2015, 35, 43). Over ten thousand people are employed by social economy organizations, representing about 20 per cent of all employment in the region (45). Co-operatives in the North play a major role in producing and selling artworks, particularly Inuit prints and sculptures, and in sponsoring and supporting cultural events, such as music, dance, drumming performances, and the Arctic Winter Games. They are often viewed as the best vehicle for maintaining local control over such cultural activities and "joining the two activities of money and culture" (MacPherson 2015, 153–4). In 2011, the tourism industry contributed $147 million to the economies of the three territories (HLT Advisory, Tourism Industry Association of Canada, and Visa Canada 2012), with cruise tourism introducing a new source of revenue.

Evolution of Cruise Tourism in Canada's North

Historically, because of its remoteness and cold climate, Canada's North was not a major centre of maritime activity, and the Canadian government's policies tended to focus mainly on the issue of Canada's sovereignty over the lands and waters that make up the Arctic Archipelago. However, despite the inhospitable climate, the high Arctic has for centuries attracted explorers and adventurers searching for the fabled Northwest Passage, a sea route that would shorten the distance between Europe and Asia by as much as ten thousand kilometres as compared to conventional routes through the Panama or Suez Canals (Guy and Lasserre 2016, 1). The most famous of these explorers—Sir John Franklin—led a doomed expedition into the Passage in 1845, where he and his entire crew of 129 men perished. It was only in 2014 and 2016 that the wrecks of his two ships—the *HMS Erebus* and the *HMS Terror*—were discovered in the waters off King William Island (Neatby and Mercer 2016).

The Norwegian explorer Roald Amundsen was the first to navigate the entire length of the approximately 7,000-kilometre-long Passage in the *Gjoa* in 1903–5, but no other ship was able to repeat this feat until 1940–2, when a Royal Canadian Mounted Police schooner, the *St. Roch*, made the trip (Headland 2020, 3). Between those two voyages and the 1970s, there were only nine transits of the Northwest Passage, mostly by icebreakers and research vessels (11). Due to the rapid decrease in ice cover as a result of climate change,[2] the number of transits increased sharply from about four per year in the 1980s to about twenty to thirty between 2009 and 2013 (Government of the Northwest Territories 2015). Since 2006, the southern route[3] of the Northwest Passage has been free of sea ice for most of the summer. By August 2016, both the northern and southern routes were almost completely ice free, and Arctic sea ice coverage was the fourth-smallest ever recorded by satellites (Di Liberto 2016). In 2019, sea ice coverage in the eastern Arctic was below 5 per cent by September, the lowest on record, leading to a record high of 147 ships visiting Canadian Arctic waters between May and October, a 70 per cent increase since 2009 (Friedman 2020).

The touristic appeal of these remote and treacherous waters was proven in 1984, when the *Lindblad Explorer* became the first cruise ship to traverse the Passage. This venture was followed by many others, and by 2015, cruise ships and cruise ship icebreakers constituted 16 per cent of the marine traffic through the Passage (Government of the Northwest Territories 2015). Between 2005 and 2014, 26 cruise ships or touristic icebreakers transited the Northwest Passage, and a total of 149 cruise ships visited the Canadian Arctic (although not all traversed the Northwest Passage), accounting for approximately 6 per cent of the total marine traffic in the Arctic in those years (Guy and Lasserre 2016, 11).

As table 5.1 indicates, as the ice has melted, the volume of cruise tourism has steadily increased. In 2016, 11 cruise ships registered with the Nunavut Department of Economic Development and Transportation, some stopping more than once in several communities. For example, in Pond Inlet, a community at the eastern approach to the Northwest Passage, 6 cruise ships visited once between August 8 and September 4, and 3 visited twice (Government of Nunavut 2016).[4] In the summer of 2018, 12 cruise ships carrying between 132 and 530 passengers registered with the Government of Nunavut, stating their intention to make 63 visits to 13 communities (DEDT 2019), although about half of these visits were cancelled due to adverse ice and weather conditions (Tranter 2019). In 2019, the Canadian Coast Guard reported that 27 vessels made full transits of the Northwest Passage (Canadian Coast Guard 2019).

While the volume is increasing, the number of cruise passengers visiting Canada's North is still far below that of other northern regions. In 2015, Alaska welcomed over a million cruise passengers, Svalbard in Norway around 35,000, and Greenland about 20,000. This compares to the roughly 4,700 cruise passengers who landed in Nunavut in 2016 (DEDT 2016). In many areas—marine support infrastructure, regulations, and tourism promotion—Canada is playing catch-up (Dawson et al. 2016a). However, interest in shipping through the Northwest Passage has steadily grown, and Canada's historical claim to sovereignty over the Passage is being challenged by the United States, Russia, and China, which see the route's

Table 5.1. Nunavut Cruise Tourism Statistics

YEAR	2010	2011	2012	2013	2014	2015	2016
Number of operators	6	4	6	6	7	8	8
Total vessels	7	5	6	8	8	11	11
Total voyages	13	9	13	21	15	23	25
Total community visits*	32	25	31	45	39	52	59
Total passengers**	1,398	1,353	2,153	3,289	1,905	3,364	4,758

Source: Government of Nunavut (quoted in Sevunts 2017)

* Community visits are based on provided itineraries and do not account for cancellations
** Passenger numbers assume maximum capacity of cruise vessels

commercial possibilities, regardless of the environmental and logistical challenges of sending ships through the area (McCoy 2016; Struzik 2016). This is creating more pressure to address the issues surrounding marine shipping, including cruises.

Cruise Tourism: Weighing the Tangible and Intangible Risks and Benefits

There are many reasons why cruise ship tourists might be interested in visiting Canada's North, but one major motivation is the notion of "last chance tourism," which has been defined as "tourists explicitly seek[ing] vanishing landscapes or seascapes, and/or disappearing natural and/or social heritage" (Lemelin et al. 2010, 478). This motivation has also been referred to as "extinction tourism" (Migdal 2016; McKie 2016) or "last frontier tourism" (Johnston et al. 2012). Whatever the terminology used, because of climate change, the Canadian Arctic's environmental, social, economic, and cultural ecosystems are under increasing stress, making the area prime ground for the "last chance" traveller (Stewart et al. 2007; Dawson et al. 2011). The tangible impacts of cruise tourism fall mainly within the environmental and economic spheres of northern communities, while the intangible ones are seen most readily in the social and cultural spheres. While the risk analysis presented below is separated into four categories, these categories are interlinked, rather than discrete, and tangible and intangible impacts often intersect categorical boundaries. As Dawson, Maher, and Slocombe (2007) have argued, a systems approach to tourism management in the Arctic is essential to understanding the interconnected impacts on biophysical and human systems.

Environmental

The most serious risk from cruise tourism stems from climate change in the Arctic, which has contributed to the rapid melting of sea ice. In assessing the environmental risks to shipping in the North, the insurance company Allianz highlighted the lack of reliable marine maps for about 90 per cent of the Arctic, the lack of rescue and oil clean-up capacity in the area if something should go wrong,[5] and the lack of training in northern navigation for many crews (Allianz n.d.). An example of these risks occurred in August 2010 when a small cruise ship, the *MV Clipper Adventurer*, carrying about two hundred passengers, ran aground near Kugluktuk, Nunavut. While there was no loss of life or oil as a result of this incident, it took more than two days for the nearest Canadian Coast Guard vessel to arrive at the scene to provide assistance (Leblanc 2016).

All types of Arctic shipping pose threats to the region's wildlife. These threats include ship strikes of marine mammals, the introduction of alien species, disruption of migratory patterns of marine mammals, toxin and noise pollution of habitat, and acceleration of ice melts from ships' carbon emissions (Arctic Council 2009; Hoag 2016; Stewart et al. 2013). Sea ice loss is also affecting the population health and geographic distribution of wildlife in the North and changing hunting, fishing, and gathering practices. Traditional travel routes have been disrupted; some areas have become inaccessible; and pressure has been put on the mixed economy of the Arctic (Dawson, Maher, and Slocombe 2007). The Traditional Knowledge of Elders is also becoming less reliable as weather patterns change. Heritage sites that had been preserved by the Arctic permafrost are becoming more vulnerable as the permafrost thaws and exposes archaeological sites and artifacts to rot and decay (Nunavut Climate Change Centre n.d.). All of these elements pose ethical dilemmas for those involved in the promotion and operation of cruise tourism in the North. To what extent can the impact of cruise tourism on an already stressed environment be justified by the economic opportunities that this activity can potentially provide to communities in the North?

Economic

Culture plays a significant role in the northern economy. In 2006, artists in the three northern territories comprised 1.02 per cent of the labour force, as compared to the national average of 0.77 per cent. In the community of Cape Dorset in Nunavut, artists made up over 9 per cent of the labour force (Hill Strategies 2010). There are an estimated 13,650 Inuit visual artists in Canada, and in 2015 they earned about $33 million from their art and generated an estimated additional $12.6 million in spin-off economic activity (Big River Analytics 2017).[6] Almost one-third of Indigenous people in the Arctic earn income from selling artworks, and some 18 per cent of Arctic residents manufacture crafts for sale (Petrov 2014, 168).

Before the *Crystal Serenity* made its historic voyage through the Northwest Passage, residents of Ulukhaktok, Cambridge Bay, and Pond Inlet, where the ship was scheduled to stop, were optimistic about cruise passenger spending on local arts and crafts (Brown 2016; Kyle 2016; Hopper 2016). However, the Nunavut Arts and Crafts Association estimates that the average cruise ship passenger spent only $75 on artworks (Hopper 2016), and the overall economic benefits to communities tended to be less than anticipated.

Cruise ship passengers spend an average of $692 in the territory, as compared to about $2,500 spent by land-based tourists, who must also pay for food and lodging (Sorensen 2016). The Nunavut government's tourism and cultural industries director has remarked that "the communities didn't feel like they were part of the economic development. They felt the tourists on the cruise ships just came, took some pictures, and left" (Murray 2016).

One of the greatest barriers to sales is the prohibition against tourists from the United States and Europe bringing home artifacts made from traditional materials such as sealskin and narwhal ivory. Importing sealskins into the United States has been illegal since 1972, and the European Union has banned sealskin products since 2009, although in 2015 the Government of Nunavut negotiated an exemption to the ban for products harvested by Indigenous Peoples (Zerehi 2016). Since 85 per cent of the tourists aboard the *Crystal Serenity* were Americans, artists lost thousands of dollars in sales. In this case, the economic outcome can be linked to what Nunavut's senior advisor on tourism and legislation has described as southern tourists' discomfort with "seeing a bloody beach full of seal carcasses" (Murray 2016).

Social

Cruise tourism has many other impacts on small Arctic communities beyond the economy. Most communities in the North have between three hundred and three thousand residents, and frequent short visits by large numbers of cruise passengers can be extremely disruptive to daily life. As one local tour operator said, "there's a lot of employment opportunities in the community when the cruise ships come in," but "for us here, because we have such minimal resources, we need at least a day between each ship" (Kassam 2016). The chair of the Inuit Circumpolar Council remarked that "far too many people will be descending suddenly into the communities and bringing far too much garbage with them" (McKie 2016). Some communities have had to hire security staff to prevent theft, and others have found that cruise tourists going online at community access centres can deplete a month's worth of bandwidth in a single day (George 2016). In one case, a ship bought all the milk and fresh produce in the communities, leaving nothing for locals until the next provisions flight from the South (Bramham 2016). People in some communities were also concerned about the possibility of ship passengers bringing illness into the community and about the potential for criminal activities such as drug and alcohol smuggling (Dawson, Johnston, and Stewart 2012d).

Decision-makers and regulators in Nunavut also worry about the lack of collaboration among communities to develop a diversity of tourism products and activities, the lack of guidelines for tour operators on how to conduct socially and culturally friendly visits, and the lack of capacity in many small communities to handle the increasing volume of cruise tourism (Johnston et al. 2012).

Residents of Arctic communities do not, however, view the social impacts of cruise tourism in a completely negative light. Interviews in several northern communities conducted by Dawson, Johnston, and

Stewart revealed that residents also welcomed the opportunity to share Inuit culture and traditions, meet new people, make friends, and reinforce a sense of pride in their communities (2012a, 2012b, 2012c). Some communities, such as Ulukhaktok, developed extensive cultural programs for the *Crystal Serenity*'s passengers, which included arts-and-crafts demonstrations, guided tours, and local foods. As one Arctic-style dancer in the community stated, "It's very exciting to show how we live here. . . . That is so good" (Anselmi 2016).

Cultural

Definitions of "culture" and "cultural tourism" used by tourism promoters can have varied meanings. In an environment such as the Canadian North, specific issues related to Indigenous cultural tourism must be added to this mix. Indigenous cultural tourism often involves visits to remote and fragile locations where environmental issues "impact greatly on the lifestyles and traditions of indigenous peoples" and their relationships with the land (Smith 2016, 129). Even in cultural domains that have been commoditized, such as the production and sale of arts and crafts, risks can accompany the benefits. In the Canadian Arctic, all these elements of culture are at play.

As the Nunavut tourism agency points out on its website, in English, "culture" has over 160 meanings, but in Inuktitut (the Inuit language) there is no such word The closest term is *illiqusiq*, which means "the way it is done." It encompasses all aspects of life, and it is worth quoting Nunavut Tourism's explanation in full to gain an appreciation of what this means:

> Inuit culture includes their language, traditions, beliefs, music, art, handicrafts, foods, clothing, implements, technologies, and story. The kayak, the ulu knife, the igloo, and the inuksuk are distinctive examples. Dog sleds are still popular in Nunavut, but snowmobiles are more common. The rifle has replaced the bow and arrow, but to the Inuit way of seeing things, this is still "traditional" because it's logical and practical. From the cultural perspective of a hunting and fishing people, using GPS to find one's way back home is as basic as replacing stone arrowheads with high-calibre rifle ammunition.

Likewise, in music, the traditional sounds of throat-singing are sometimes now mixed to hip hop beats. Storytelling, which is a traditional performance art form, nowadays also includes the innovative work of Inuit filmmakers. The Inuit culture that people will experience when visiting Nunavut today is both vibrant and dynamic. It is an ancient, living culture (Nunavut Tourism n.d.b).

Although northerners have been remarkably open to change and innovation, as Smith observes, "it is an inevitable fact of tourism that cultural changes occur primarily to the indigenous society's traditions, customs and values rather than to those of the tourist" (2016, 237), and this concern also applies to tourists' often "selective" interest in Indigenous artistic expression (142). Threats to authenticity were voiced as far back as the 1990s. A survey of the residents of Cape Dorset, where Inuit carvings and prints for the commercial art market have been produced since the 1950s, found that "27% of respondents referred to . . . loss of artifacts from the community, negative cultural impacts such as the debasing of local art, and the potential for the community to lose control over the industry's development" (Milne, Ward, and Wenzel 1995, 30).

More recent surveys have found similar unease with the cultural impacts of cruise tourism, although this was often counterbalanced by residents' desire to showcase their culture and share it with visitors. Some residents identified cultural risks, such as a sense of intrusion by cruise visitors, with limited understanding of subsistence lifestyles, inappropriate photography of local people and property, and disturbance of historically or culturally significant sites (Stewart, Dawson, and Johnston 2015). However, other residents believed that cruise tourism helped to keep traditional drum dancing and singing alive and encouraged Inuit "to express our cultural knowledge and show how our ancestors used to play" (413).

The creative aspects of such tourism are often viewed as opportunities to pass along knowledge to the younger generation and to work together to share local history with visitors. For some residents, "this collaborative approach to showcasing their community to visitors was regarded as one of the key benefits of cruise ship visitors," contributing to pride in the culture (Stewart, Dawson, and Johnston 2015, 413). In some cases, cruise tourism has also encouraged performers to expand their repertoire. In Pond Inlet, a community that sees multiple cruise ships each summer, one resident indicated that performers were now working "to make the show a little more exciting—they used to sing 'ayayah' [songs from the ancestors] but it seemed a little boring for the people, so . . . they made up a little play as well" (413).

Overall, cultural programming in the Arctic strives to be real and not captive to what Smith terms "the romanticised" or the "exotic" (2016, 140). While cruise tourism builds interest in the heritage and traditions of the northern Indigenous Peoples, a delicate balance is needed to ensure that this interest does not overwhelm communities and that it contributes to, rather than erodes, sustainability. As Pelly argues:

> The cultural experience offered to visitors must be a genuine reflection of the community's life, with Inuit participants doing what they really do and visitors invited to share in that experience. What this means is simply that cultural programs must be developed for the community's sake, not just for the tourists. . . . The visitor who is invited to share in the experience, even as a detached observer, is witnessing a living part of the community's cultural reality. If the cultural programs are real, are alive, then the "cultural tourism" will be sustainable. (2013, iii)

Indigenous World Views and Planning Frameworks: Assessing the Adaptive Capacities of Northern Communities

Sustainability is a governance challenge that must take into account not only the world views of northern residents, but also the legal and administrative frameworks that have been overlaid on them by federal, territorial, and local governments. Because of the nature of the Canadian federation, issues such as climate change, marine shipping and safety, and tourism development do not fall within neat jurisdictional categories, but are, rather, subject to varying degrees of multi-level governance (Higginbotham 2013; Rodon 2015). Jurisdictionally, the Government of Canada has the primary responsibility for marine policies but the Government of Nunavut is responsible for land-use planning, which seeks "to find a broadly acceptable compromise between protecting Nunavut's pristine land and water, and allowing tourism, mining, and other types of development" (Kujawinski 2017).

As a result, small communities in the North cannot respond autonomously to the pressures of cruise tourism. Not only are they required to work with various arms of the federal government, such as the Canadian Coast Guard, Transport Canada, and Environment Canada, but also the territorial departments of environment, economic development, and transportation, as well as industry associations like the Association of Arctic Expedition Cruise Operators (AECO), and various international bodies, such as the International Maritime Organization. Meshing the priorities of all these institutional players with the needs and requirements of residents in northern communities is therefore a complicated, but necessary, process.

Assessing the response of northern communities to the risks and opportunities of cruise tourism requires an understanding of the way in which Indigenous world views are integrated into the planning and regulatory frameworks of the region. Jeannotte (2017) has argued that holistic world views, such as that of the medicine wheel, are central to Indigenous Peoples' relationship to the land, as well as to their identities and well-being, but have only recently been incorporated into local sustainability planning in a limited way. Matunga (2013) notes that place-based planning in many Indigenous settings has been misappropriated by colonial powers, but that common themes of Indigenous decision-making include a determination to strive for consensus, use core values and Traditional Knowledge to guide the decision-making process, incorporate the wisdom of Elders, and conduct planning processes and meetings according to cultural protocols.

Recognizing the need to incorporate Indigenous world views into its public administration, the Government of Nunavut has directed that the Inuit concept of *Qaujimajatuqangit* be taken into consideration in all areas of policy development (Wenzel 2004). Inuit *Qaujimajatuqangit* is a unified system of beliefs founded on four principles: working for the common good; respecting all living things; maintaining harmony and balance; and continually planning and preparing for the future (Tagalik 2009, 1). It forms the basis of many planning processes in Nunavut, including the *Iqaluit Sustainable Community Plan* (2014), which states that Inuit *Qaujimajatuqangit*:

> encompasses all aspects of traditional and modern Inuit culture, including wisdom, behaviours, world-view, beliefs, language, relationships, life skills, perceptions, and expectations. Inuit *Qaujimajatuqangit* helps us to better understand and adapt to today's changes and challenges. It recognizes that everything is related to everything else, in such a way that nothing can stand alone. This is actually the pulse of our sustainability (4).

Traditional governance structures in Inuit communities are inclusive and consensus-based (Pauktuutit 2006; Ritsema et al. 2015). However, contemporary governance systems in the North are rooted in Weberian-style hierarchical bureaucracies. Ritsema et al. have noted that such governance styles are often at odds with traditional community approaches. They have suggested that a better cultural match is needed to achieve a "more acceptable balance between Western and Aboriginal belief systems, modern and traditional lifestyles, Canadian and Inuit governance, and among different age groups in Inuit society" (2015, 170).

The need to bring culture into the planning mix is evident in a study of cruise tourism conducted by Dawson, Johnston, and Stewart (2012d), which included 270 interviews with local residents of seven Arctic communities. The study also surveyed 42 policy-makers and regulators, as well as 18 cruise operators. Concerns with safety, economic development, social interactions, culture, and the environment were identified. Strategies to deal with the cultural issues included the creation of tourism codes of conduct, marketing training for local artists, cultural sensitivity training for cruise tourists, and enhanced education for tourists about traditional and contemporary Inuit lifestyles.

Governments and industry associations in the North are taking heed of these concerns and are attempting to adapt program and planning initiatives to address them. The Government of Nunavut's Department of Economic Development and Transportation (DEDT) has developed a *Nunavut Marine Tourism Management Plan 2016–2019* that includes activities in seven areas:

- Planning with communities
- Understanding economic impacts
- Helping communities prepare
- Providing information and resources
- Providing input for development support
- Developing and implementing regulations and policy supports
- Communicating with industry and visitors (DEDT 2016, 8)

It also states that "For marine tourism development to be successful, it must contribute to the economic well-being of individuals, businesses, and communities without negative social and cultural outcomes and it must not exhaust local and territorial government management resources needed to organize and control it" (12). Moreover, it has expanded the capacity of the government's Community Tourism and Cultural Industries Program to provide assistance for arts creation, tourism development, tourism and arts marketing, and development of new arts, culture, and tourism infrastructure (DEDT n.d.). In addition, the AECO began work in 2017 to develop new guidelines of behaviour for cruise ship visitors and host communities (Nunatsiaq News 2017). In 2018, the AECO signed a memorandum of understanding with the Nunavut DEDT, forming a Community Engagement Committee to collaborate in a number of key areas, such as training and guidelines for visits to archaeological sites (DEDT 2019, 29–30). Since January 2017, cruise companies sailing in the Canadian Arctic are also subject to the Polar Code, developed by the UN's International Marine Organization (Sevunts 2017).

Ship traffic in the Arctic tends to coincide with areas that Inuit and other stakeholders have identified as "sensitive cultural sites," which often correlate with sensitive ecological sites (Dawson et al. 2016b, 7–8). Recently, the Government of Nunavut has responded by conveying community concerns about ships disturbing traditional harvesting areas directly to cruise operators and by monitoring incidents as they occur (DEDT 2019, 31). However, Dawson and her colleagues point to the broader need "to identify and prioritize existing local and cultural risks, including to culture, lifestyle, wildlife, and the local environment" (6). In a separate analysis of the policy and governance challenges, Stewart, Dawson, and Johnston have concluded that the vast geographical territory and decentralized management of Arctic cruise tourism has caused "a diffusion of responsibility among large numbers of organizations and departments, leading to management gaps, oversights, and communication difficulties" (2016, 27). In view of such inconsistencies, the same scholars have recommended that a harmonized policy framework be "established to ensure environmental and human risk is minimized, and economic and cultural opportunities are maximized" (30).

Conclusion

Culture is an integral part of the Inuit way of life. Therefore, adjustments to the consequences of cruise tourism in the North must also include cultural considerations and must proceed from a fundamental concern about the sustainability of life in the region. To this end, several factors must come together to ensure that cruise tourism is integrated within a sustainable framework in Canada's North:

- Canada should become more proactive in integrating all cruise tourism stakeholders.

- The predominant economic model in the North—a mixed economy with a strong social economy element—should be harnessed more consistently to harvest the benefits of cruise tourism.

- Indigenous world views and the traditional resilience of northern communities should be taken into consideration when integrating cultural resources into planning and regulatory frameworks for managing cruise tourism.

With regard to the first point, the territorial governments, especially Nunavut's, and local communities are attempting to address the issues that fall within their jurisdictions, but the Canadian government still appears to be working in a number of silos with regard to safety, security, monitoring of traffic, infrastructure construction and maintenance, and tourism promotion. As Dawson and her colleagues observe, addressing the challenges of cruise tourism in the North will "demand a great deal of political will and determination at all levels of government, within all sectors, and amongst all stakeholder groups" (Dawson et al. 2016a, 1438). This is unlikely to happen unless concerns about Canadian sovereignty in the North are combined with environmental and commercial pressures to overcome bureaucratic inertia within the higher levels of government.

Social economy organizations in the North are active in both the tourism and arts-and-culture fields, and they are increasingly likely to be relied upon in the area of cruise tourism. For example, in the community of Arctic Bay, the Arctic Bay Adventures Co-op is owned by the community and proceeds from tourism are shared within the community (Sevunts 2017). This type of economic structure ensures that communities benefit more than outside businesses from the visits of cruise tourists, and allows local residents to control both the quantity and quality of tourist activities. Given the vastness of the region and the place-based nature of many of the environmental and cultural risks described above, community residents are better placed than other stakeholders to foster a sustainable cruise tourism industry into the future.

This chapter has discussed how culture as way of life, as vehicle for sustainable values, and as complex network of social and economic systems forms the basis of much of daily life in the North. It stands to reason that these frameworks should also apply when planning for and managing culture as capital (in both its tangible and intangible forms) and as forms of creative expression (including crafts, works of art, and performance). Considering that northern societies are built upon traditional values of sharing and community relationships, the economic aspects of cruise tourism in Canada's Arctic cannot readily be severed from its social, cultural, and environmental elements without compromising the cohesive fabric of local communities. When the cruise ships do come into these communities, they are not simply vehicles for their passengers. They are becoming part of the lives and relational fabrics of northern communities, with all the complex cultural interactions that this implies.

NOTES

1   Three Aboriginal groups are recognized in the Canadian Constitution—First Nations, Inuit, and Métis (people of mixed European and Aboriginal heritage). The majority of Inuit are located in Nunavut, although there are small numbers in the Northwest Territories, Nunavik (northern Quebec), and Nunatsiavut (Newfoundland and Labrador). First Nations and Métis Peoples in the North are primarily located in the Yukon and the Northwest Territories.

2   Satellite data show that since 1979, the summer sea ice has retreated by an average of 13.7 per cent per decade (Guy and Lasserre 2016, 6).

3   There are several routes through the Northwest Passage, owing to the many islands and straits in the Arctic Archipelago.

4   Despite the summer retreat of the ice, Arctic waters are only navigable by cruise ships in a narrow window of time between late July and early September.

5   Canada has only fifteen icebreakers to patrol the longest coastline in the world. In 2016, only seven were assigned to the Canadian Arctic, an area spanning 4.4 million square kilometres (Sorenson 2016).

6   Figures here and below are in Canadian dollars.

References

Allianz. n.d. "Shipping and the Northwest Passage—Future Growth and Risks." Accessed 19 October 2020. https://insurancemarinenews.com/insurance-marine-news/shipping-northwest-passage-future-growth-risks/.

Anselmi, Elaine. 2016. "On Land and Water, Ulukhaktok Welcomes Luxury Cruiseliner, *Arctic Deeply*." *New Humanitarian*, 8 September. https://www.newsdeeply.com/arctic/articles/2016/09/08/on-land-and-water-ulukhaktok-welcomes-luxury-cruiseliner.

Arctic Council. 2009. *Arctic Marine Shipping Assessment 2009 Report*. Accessed 1 June 2017. https://www.pmel.noaa.gov/arctic-zone/detect/documents/AMSA_2009_Report_2nd_print.pdf.

Bennett, Nathan, and Harvey Lemelin. 2015. "Conservation-cum Social and Economic Development: The Emergence of an Eco-social Economy in the Canadian North." In *Northern Communities Working Together: The Social Economy of Canada's North*, edited by C. Southcott, 228–52. Toronto: University of Toronto Press.

Big River Analytics. 2017. *The Impact of the Inuit Arts Economy*. Ottawa: Indigenous and Northern Affairs Canada. Accessed 19 October 2020. https://www.rcaanc-cirnac.gc.ca/eng/1499360279403/1534786167549.

Bramham, Daphne. 2016. "Arctic Region Must Adapt to Less Ice, but Can It Happen Fast Enough?" *Vancouver Sun*, 3 September. http://vancouversun.com/opinion/columnists/daphne-bramham-arctic-region-must-adapt-to-less-ice-but-can-it-happen-fast-enough.

Brown, Chris. 2016. "Massive Cruise Ship Brings New Era of Arctic Tourism to Cambridge Bay—Visit Was Encouraged by Artisans Who Stand to Make Thousands of Dollars in Sales in 1 Day." *CBC News*, 29 August. http://www.cbc.ca/news/canada/north/massive-cruise-ship-brings-new-era-of-arctic-tourism-to-cambridge-bay-1.3739491.

Butler, Richard. 2015. "Sustainable Tourism—Paradoxes, Inconsistencies and a Way Forward?" In *The Practice of Sustainable Tourism: Resolving the Paradox*, edited by Michael Hughes, David Weaver, and Christof Pforr, 66–79. New York: Routledge.

Canadian Coast Guard. 2019. "2019 Arctic Operations for Canadian Coast Guard Complete." News release, 2 December. https://www.canada.ca/en/canadian-coast-guard/news/2019/12/2019-arctic-operations-for-the-canadian-coast-guard-complete.html.

Canadian Northern Economic Development Agency. 2016. *Northern Economic Index*. Accessed 18 May 2017. http://www.cannor.gc.ca/eng/1387900596709/1387900617810.

Conference Board of Canada. 2014. "Education and Skills in the Territories" *How Canada Performs*. Accessed 18 May 2017. http://www.conferenceboard.ca/hcp/provincial/education/edu-territories.aspx.

Dawson, Jackie, Margaret Johnston, and Emma Stewart. 2012a. *Cruise Tourism in Arctic Canada—Community Report for Gjoa Haven*. Accessed 19 October 2020. https://www.espg.ca/wp-content/uploads/2013/04/3-C-TAC_Gjoa-Haven_En.pdf.

———. 2012b. *Cruise Tourism in Arctic Canada—Community Report for Pond Inlet*. Accessed 19 October 2020. https://www.espg.ca/wp-content/uploads/2013/04/5-C-TAC_Pond-Inlet_En.pdf.

———. 2012c. *Cruise Tourism in Arctic Canada—Community Report for Ulukhaktok*. Accessed 24 October 2016. http://artsites.uottawa.ca/ctac/.

———. 2012d. *Cruise Tourism in Arctic Canada—Final Report*. Accessed 19 October 2020. https://www.espg.ca/wp-content/uploads/2013/12/C-TAC_Final-Report_En.pdf.

Dawson, Jackie, Margaret Johnston, Emma Stewart, C. J. Lemieux, Harvey Lemelin, Patrick Maher, and B. S. R. Grimwood. 2011. "Ethical Considerations of Last Chance Tourism." *Journal of Ecotourism* 10, no. 3: 250–6.

Dawson, Jackie, Patrick Maher, and Scott Slocombe. 2007. "Climate Change, Marine Tourism, and Sustainability in the Canadian Arctic: Contributions from Systems and Complexity Approaches." *Tourism in Marine Environments* 4, nos. 2–3: 69–83.

Dawson, Jackie, Emma Stewart, Margaret Johnston, and Christopher Lemieux. 2016a. "Identifying and Evaluating Adaptation Strategies for Cruise Tourism in Arctic Canada." *Journal of Sustainable Tourism* 24, no. 10: 1425–41.

Dawson, Jackie, Louie Porta, Seyi Okuribido-Malcolm, Maggie deHann, and Olivia Mussells. 2016b. *Proceedings of the Northern Marine Transportation Corridors Workshop, December 8.* Vancouver and Ottawa: uO Research.

DEDT (Department of Economic Development and Transportation). 2016. *Nunavut Marine Tourism Management Plan 2016–2019.* Iqaluit: Government of Nunavut.

———. 2019. *Annual Tourism Report 2018–2019.* Iqaluit: Government of Nunavut.

———. n.d. "Community Tourism and Cultural Industries Program." Accessed 28 June 2017. http://gov.nu.ca/edt/programs-services/community-tourism-and-cultural-industries-program.

Di Liberto, Tom. 2016. "Northwest Passage Clear of Ice Again in 2016." National Oceanic and Atmospheric Administration's Climate.gov website. Accessed 9 February 2017. https://www.climate.gov/news-features/event-tracker/northwest-passage-clear-ice-again-2016.

Duxbury, Nancy, and M. Sharon Jeannotte. 2013. "The Role of Cultural Resources in Community Sustainability: Linking Concepts to Practice and Planning." *International Journal of Sustainability Policy and Practice* 8, no. 4: 133–44.

Friedman, Gabriel. 2020. "Climate Change Offers Invitation to Explore the Arctic as Marine Traffic Hits Unprecedented Levels." *Financial Post,* 2 January. https://business.financialpost.com/commodities/mining/climate-change-offers-invitation-to-explore-the-arctic-as-marine-traffic-hits-unprecedented-levels.

George, Jane. 2016. "Nunavut's Northwest Passage Gateway Braces for Giant Cruise Ship." *NunatsiaqOnline,* 15 August. http://www.nunatsiaqonline.ca/stories/article/65674nunavuts_nw_passage_gateway_cambridge_bay_braces_for_giant_cruise/.

Government of the Northwest Territories. 2015. "Trends in Shipping in the Northwest Passage and the Beaufort Sea." Accessed 27 January 2017. http://www.enr.gov.nt.ca/state-environment/73-trends-shipping-northwest-passage-and-beaufort-sea.

Government of Nunavut. 2016. "Master Nunavut Cruise Ship Itinerary 2016." Accessed 18 May 2017. http://www.gov.nu.ca/edt/documents-tourism.

Guy, Emmanuel, and Frédéric Lasserre. 2016. "Commercial Shipping in the Arctic: New Perspectives, Challenges and Regulations." *Polar Record* 52, no. 3: 294–304. https://doi.org/10.1017/S0032247415001011.

Haalboom, Bethany, and David Natcher. 2013. "The Power and Peril of 'Vulnerability': Lending a Cautious Eye to Community Labels." In *Reclaiming Indigenous Planning,* edited by Ryan Walker, Ted Jojola, and David Natcher, 356–75. Montreal and Kingston: McGill-Queen's University Press.

Headland, Robert K. 2020. "Transits of the Northwest Passage to End of the 2019 Navigation Season." Accessed 19 October 2020. https://www.spri.cam.ac.uk/resources/infosheets/northwestpassage.pdf.

Hicks, Jack. 2015. "Statistical Data on Death by Suicide by Nunavut Inuit, 1920 to 2014." Nunavut Tunngavik Inc. https://www.tunngavik.com/files/2015/09/2015-09-14-Statistical-Historical-Suicide-Date-Eng.pdf.

Higginbotham, John. 2013. "Nunavut and the New Arctic." *Policy Brief No. 27*. Waterloo, ON: Centre of International Governance Innovation.

Hill Strategies. 2010. "Artists in Small and Rural Municipalities in Canada—Based on the 2006 Census." https://hillstrategies.com/resource/artists-in-small-and-rural-municipalities-in-canada/.

HLT Advisory, Tourism Industry Association of Canada, and Visa Canada. 2012. *The Canadian Tourism Industry—A Special Report, Fall 2012*. http://tiac.travel/_Library/documents/The_Canadian_Tourism_Industry_-_A_Special_Report_Web_Optimized_.pdf.

Hoag, Hannah. 2016. "Ocean Warming Is Already Affecting Arctic Fish and Birds." *Arctic Deeply*, 8 September. https://www.newsdeeply.com/arctic/articles/2016/09/08/ocean-warming-is-already-affecting-arctic-fish-and-birds.

Hopper, Tristin. 2016. " 'Do You Live Here All Year?' Nunavut Community Invaded by Largest Cruise Ship in Arctic History." *National Post*, 29 August. http://news.nationalpost.com/news/do-you-live-here-all-year-nunavut-community-invaded-by-largest-cruise-ship-in-arctic-history.

Jeannotte, M. Sharon. 2017. "Caretakers of the Earth: Integrating Canadian Aboriginal Perspectives on Culture and Sustainability into Local Plans." *International Journal of Cultural Policy* 23, no.2: 199–213.

Johnston, Adrianne, Margaret Johnston, Emma Stewart, Jackie Dawson, and Harvey Lemelin. 2012. "Perspectives of Decision Makers and Regulators on Climate Change and Adaptation in Expedition, Cruise Ship Tourism in Nunavut." *Northern Review* 35 (Spring): 69–95.

Kassam, Ashifa. 2016. "Arctic Cruise Boom Poses Conundrum for Canada's Indigenous Communities." *The Guardian*, 4 October. https://www.theguardian.com/world/2016/oct/04/arctic-cruise-boom-canada-inuit-indigenous-communities.

Kujawinski, Peter. 2017. "The Complicated Relationship between Cruise Ships and the Arctic Inuit." *The New Yorker*, 11 May. http://www.newyorker.com/news/news-desk/the-complicated-relationship-between-cruise-ships-and-the-arctic-inuit.

Kyle, Kate. 2016. "Arctic Hamlets Prepare for Giant Cruise Ship *Crystal Serenity*." *CBC News*, 17 August. http://www.cbc.ca/news/canada/north/crystal-serenity-arctic-hamlets-prepare-1.3723425.

Landriault, Mathieu. 2019. *Media, Security and Sovereignty in the Canadian Arctic: From the Manhattan to the Crystal Serenity*. London: Routledge. https://www.taylorfrancis.com/books/9780367816292.

Leblanc, Pierre. 2016. "Canadian Arctic Is No Place for Large Cruise Ships." *Arctic Deeply*, 20 April. https://www.newsdeeply.com/arctic/community/2016/04/20/canadian-arctic-is-no-place-for-large-cruise-ships.

Lemelin, Harvey, Jackie Dawson, Emma Stewart, Patrick Maher, and Michael Lück. 2010. "Last Chance Tourism: The Doom, the Gloom, and the Boom of Visiting Vanishing Destinations." *Current Issues in Tourism* 13, no. 5: 477–93.

Lemelin, Harvey, Margaret Johnston, Jackie Dawson, Emma Stewart, and Charles Mattina. 2012. "From Hunting and Fishing to Cultural Tourism and Ecotourism: Examining the Transitioning Tourism Industry in Nunavik." *Polar Journal* 2, no. 1: 39–60.

MacPherson, Ian. 2015. "Beyond Their Most Obvious Face: The Reach of Cooperatives in the Canadian North." In *Northern Communities Working Together: The Social Economy of Canada's North*, edited by Chris Southcott, 139–61. Toronto: University of Toronto Press.

Matunga, Hirini. 2013. "Theorizing Indigenous Planning." In *Reclaiming Indigenous Planning*, edited by Ryan Walker, Ted Jojola, and David Natcher, 3–32. Montreal and Kingston: McGill-Queen's University Press.

McCoy, Jeremy. 2016. "Northwest Passage: Trump Card for US Arctic Policy?" *Global Research*, 7 December. http://www.globalresearch.ca/northwest-passage-trump-card-for-us-arctic-policy/5560748.

McKie, Robin. 2016. "Inuit Fear They Will Be Overwhelmed as 'Extinction Tourism' Descends on Arctic." *The Guardian*, 21 August. https://www.theguardian.com/world/2016/aug/20/inuit-arctic-ecosystem-extinction-tourism-crystal-serenity.

Migdal, Alex. 2016. "Cruise Ship Looks to Make Clean Journey through Northwest Passage." *The Globe and Mail*, 19 August. https://www.theglobeandmail.com/news/national/cruise-ship-looks-to-make-clean-journey-through-northwest-passage/article31478675/.

Milne, S., S. Ward, and G.Wenzel. 1995. "Linking Tourism and Art in Canada's Eastern Arctic: The Case of Cape Dorset." *Polar Record* 31, no. 176: 25–36.

Murray, Nick. 2016. "Spend! Spend! Nunavut Comes up with Plan for Growing Cruise Ship Traffic." *CBC News*, 17 June. http://www.cbc.ca/news/canada/north/nunavut-cruise-ship-management-1.3639905.

Neatby, Leslie H., and Keith Mercer. 2016. "Sir John Franklin." *Canadian Encyclopedia*. http://www.thecanadianencyclopedia.ca/en/article/sir-john-franklin/.

Nordicity Group and Uqsiq Communications. 2010. *Economic Impact Study: Nunavut Arts and Crafts—Final Report*. Iqaluit: Nunavut Department of Economic Development and Transportation.

Nunatsiaq News. 2017. "Arctic Cruise Operators Promote New Guidelines for Passengers, Communities." *NunatsiaqOnline.ca*, 16 May. http://www.nunatsiaqonline.ca/stories/article/65674arctic_cruise_operators_promote_new_guidelines_for_passengers_communit/.

Nunavut Climate Change Centre. n.d. "Climate Change Impacts." Accessed 17 April. http://www.climatechangenunavut.ca/en/understanding-climate-change/climate-change-impact.

Nunavut Tourism. n.d.a. "Weather & Climate." *Travel Nunavut*. Accessed 25 April 2017. http://nunavuttourism.com/about-nunavut/weather-climate.

———. n.d.b. "Arts, Crafts & Clothing." *Travel Nunavut*. Accessed 18 April 2017. http://nunavuttourism.com/things-to-see-do/arts-crafts-clothing.

Pauktuutit (Inuit Women of Canada). 2006. "The Inuit Way: A Guide to Inuit Culture." Accessed 26 June 2017. http://kugluktukhighschool.ca/the-inuit-way-a-guide-to.pdf.

Pelly, David. 2013. "Cultural Tourism in Nunavut." *Arctic* 66, no. 1 (March): iii–iv.

Petrov, Andrey. 2014. "Cultural Economy of the Arctic." In *Arctic Development Report—Regional Processes and Global Linkages*, 168. Copenhagen: Nordic Council of Ministers.

Rana, Ratna, and Awais Piracha. 2007. "Cultural Frameworks." In *Urban Crisis*, edited by M. Nadarajah and A. T. Yamamoto, 13–50. Tokyo: United Nations University Press.

Richards, Greg. 2011. "Creativity and Tourism: The State of the Art." *Annals of Tourism Research* 38, no. 4: 1225–53.

Ritsema, Roger, Jackie Dawson, Margret Jorgensen, and Brenda Macdougall. 2015. " 'Steering Our Own Ship?' An Assessment of Self-Determination and Self-Governance for Community Development in Nunavut." *Northern Review* 41: 157–80. https://doi.org/10.22584/nr41.2015.007.

Rodon, Thierry. 2015. "Land Claim Organizations and the Social Economy in Nunavut and Nunavik." In *Northern Communities Working Together: The Social Economy of Canada's North*, edited by Chris Southcott, 97–115. Toronto: University of Toronto Press.

Sevunts, Levon. 2017. "Nunavut Gears up for Increase in Arctic Tourism." *Eye on the Arctic—RCInet.ca*, 28 March 28. http://www.rcinet.ca/eye-on-the-arctic/2017/03/28/nunavut-gears-up-for-increase-in-arctic-tourism/.

Smith, Melanie. 2016. *Issues in Cultural Tourism Studies*, 3rd ed. New York: Routledge.

Sorensen, Chris. 2016. "The One Per Cent Are Coming to Canada's Arctic." *Maclean's*, 19 July. http://www.macleans.ca/economy/the-one-per-cent-are-coming-to-canadas-arctic/.

Southcott, Chris, and Valoree Walker. 2015. "A Portrait of the Social Economy of Northern Canada." In *Northern Communities Working Together: The Social Economy of Canada's North*, edited by Chris Southcott, 21–51. Toronto: University of Toronto Press.

Statistics Canada. n.d. "Population and Dwelling Count Highlight Tables, 2016 Census." Accessed 25 April 2017. http://www12.statcan.gc.ca/census-recensement/2016/dp-pd/hlt-fst/pd-pl/Table.cfm?Lang=Eng&T=101&S=50&O=A.

Stewart, Emma, Jackie Dawson, Stephen Howell, Margaret Johnston, T. Pearce, and Harvey Lemelin. 2013. "Local Level Responses to Sea Ice Change and Cruise Tourism in Arctic Canada's Northwest Passage." *Polar Geography* 36, nos. 1–2: 142–62.

Stewart, Emma, Jackie Dawson, and Margaret Johnston. 2015. "Risks and Opportunities Associated with Change in the Cruise Tourism Sector: Community Perspectives from Arctic Canada." *Polar Journal* 5, no. 2: 403–27.

———. 2016. "Growth and Challenges in Cruise Tourism in Arctic Canada." *Northern Public Affairs* 4, no. 3 (October): 27–30. https://www.northernpublicaffairs.ca/index/volume-4-issue-3/growth-and-challenges-in-cruise-tourism-in-arctic-canada/.

Stewart, Emma, Stephen Howell, Dianne Draper, J. J. Yackel, and A. Tivy. 2007. "Sea Ice in Canada's Arctic: Implications for Cruise Tourism." *Arctic* 60, no. 4: 370–80.

Struzik, Ed. 2016. "Full Speed Ahead: Shipping Plans Grow as Arctic Ice Fades." *Yale Environment 360*, 17 November. http://e360.yale.edu/features/cargo_shipping_in_the_arctic_declining_sea_ice.

Sustainable Iqaluit. 2014. *Iqaluit Sustainable Community Plan Part One—Overview*. Iqaluit: Municipal Corporation of the City of Iqaluit. https://sustainableiqaluit1.wordpress.com/resources/iqaluit-sustainable-community-plan/.

Tagalik, Shirley. 2009. *Inuit Qaujimajatuqangit: The Role of Indigenous Knowledge in Supporting Wellness in Inuit Communities in Nunavut*. Prince George, BC: National Collaborating Centre for Aboriginal Health.

Tranter, Emma. 2019. "Over 34,000 More People Visited Nunavut in 2018 than in 2015." *NunatsiaqOnline.ca*, 12 November. https://nunatsiaq.com/stories/article/over-34000-more-people-visited-nunavut-in-2018-than-in-2015/.

UNESCO. 2006. *Towards Sustainable Strategies for Creative Tourism*. Discussion Report of the Planning Meeting for 2008 International Conference on Creative Tourism, Santa Fe, New Mexico, 25–7 October 2006.

———. 2017. "UNESCO Joins Launch of International Year of Sustainable Tourism 2017." http://en.unesco.org/news/unesco-joins-launch-international-year-sustainable-tourism-2017.

UNWTO (United Nations World Tourism Organisation). 2014. "Sustainable Development of Tourism." Accessed 23 October 2017. http://sdt.unwto.org/content/about-us-5.

Verbeek, Desirée, and Hans Mommaas. 2008. "Transitions to Sustainable Tourism Mobility: The Social Practices Approach." *Journal of Sustainable Tourism* 16, no. 6: 629–44.

Weaver, Desirée. 2015. "Enlightened Mass Tourism as a 'Third Generation' Aspiration for the Twenty-First Century." In *The Practice of Sustainable Tourism: Resolving the Paradox*, edited by Michael Hughes, David Weaver, and Christof Pforr, 11–23. New York: Routledge.

Wenzel, George. 2004 "From TEK to IQ: Inuit Qaujimajatuqangit and Inuit Cultural Ecology." *Arctic Anthropology* 41, no. 2: 238–50.

Zerehi, Sima. 2016. "*Crystal Serenity* Brings Sales Boom to Nunavut Artists? Not so Fast." *CBC News*, 30 August. http://www.cbc.ca/news/canada/north/crystal-serenity-american-tourists-banned-sealskin-art-1.3741804.

Creative Tourism: The Path to a Resilient Rural Icelandic Community

Jessica Faustini Aquino, Georgette Leah Burns

Introduction

Many of the peripheral rural areas in Iceland had to undergo economic restructuring following the 2008 world economic crash, further hurting independent fishing families and small coastal communities already made fragile by the individual transferable quota system (Chambers, Helgadóttir, and Carothers 2017; Willson 2016). Transitioning from a predominately fishing economy, many of these rural areas embraced the tourism industry as a path toward economic sustainability and growth. Rural tourism development depends on a wide range of both publicly and privately owned natural and cultural resources, including food and accommodations, visitor and interpretative centres, and goods (Cawley and Gillmor 2008). Tourism is seen as part of the creative industry because of the flexibility and innovative forms of tourism needed to set particular experiences apart from others (Richards 2011). The creative person draws on their creativity and capacities to deal with economic and societal changes, which, in many ways, are an advantage as a form of survival. Innovation, action, and the capacity to deal with change distinguishes the creative person from others (Jóhannesson and Lund 2017; Richards 2011).

Creative tourism destinations offer the potential to enhance the local economy and community livelihoods by producing authentic and creative products for consumption. Over the past few decades the residents of

Húnaþing vestra proactively built a creative tourism product as a strategy for enhancing resilience in their small rural community in northwest Iceland. Using a case study analysis approach, coupled with data collected from observations and interviews, we explore creative tourism in Húnaþing vestra, describing the concept behind Selasetur Íslands (the Icelandic Seal Center) and how it continues to play an integral part in maintaining a novel approach to supporting sustainable cultural development.

Location and Cultural Context

In the ninth century, Iceland became the last country settled in Europe by people of Nordic and Celtic origin (Reynarsson 1999). The country's physical environment and many of its key historical events, such as plagues and famines, had a critical impact in shaping the life and fate of its communities. The Central Highlands cover about 40 per cent of the country and are uninhabited (Sæthórsdóttir, Hall, and Saarinen 2011), with the majority of settlements found along or near the coastline. During the early years of Icelandic settlement, people were faced with many environmental hardships; these included the effects of human settlement (such as deforestation) as well as natural disasters (such as volcanic eruptions) (Burns 2018). Such hardships were also felt by the community living in the region now encompassed by Húnaþing vestra, in northwest Iceland.

Húnaþing vestra was founded in 1998 after a merger of seven smaller municipalities, with an eighth joining in 2012. Hvammstangi, the most densely populated area, with 573 residents (Statistics Iceland 2019), is the government administrative centre for Húnaþing vestra. *Húnn* in Icelandic means polar bear cub, and Húnaþing vestra is situated in Húnaflói, or "Polar Bear Cub Bay," named after the polar bear sightings mentioned in the region's early sagas. The municipality's emblem is two polar bear cubs (see figure 6.1), while Húnavatnshreppur, the municipality directly to the east, has a mother polar bear as its emblem. Húnaþing vestra covers 3,007 square kilometres (Ráðbarður Sf., personal communication, 24 January 2020) and contains a total population of 1,210 (Statistics Iceland 2019). The cultural diversity of Húnaþing vestra has increased since 2019, such that around 11 per cent of the population now originates from seventeen non-Icelandic nationalities (municipality employee, personal communication, 23 September 2020). Not included in this count are people

Figure 6.1. Map of Húnaþing vestra (shaded region), showing the villages of Hvammstangi, Laugarbakki, and Borðeyri.

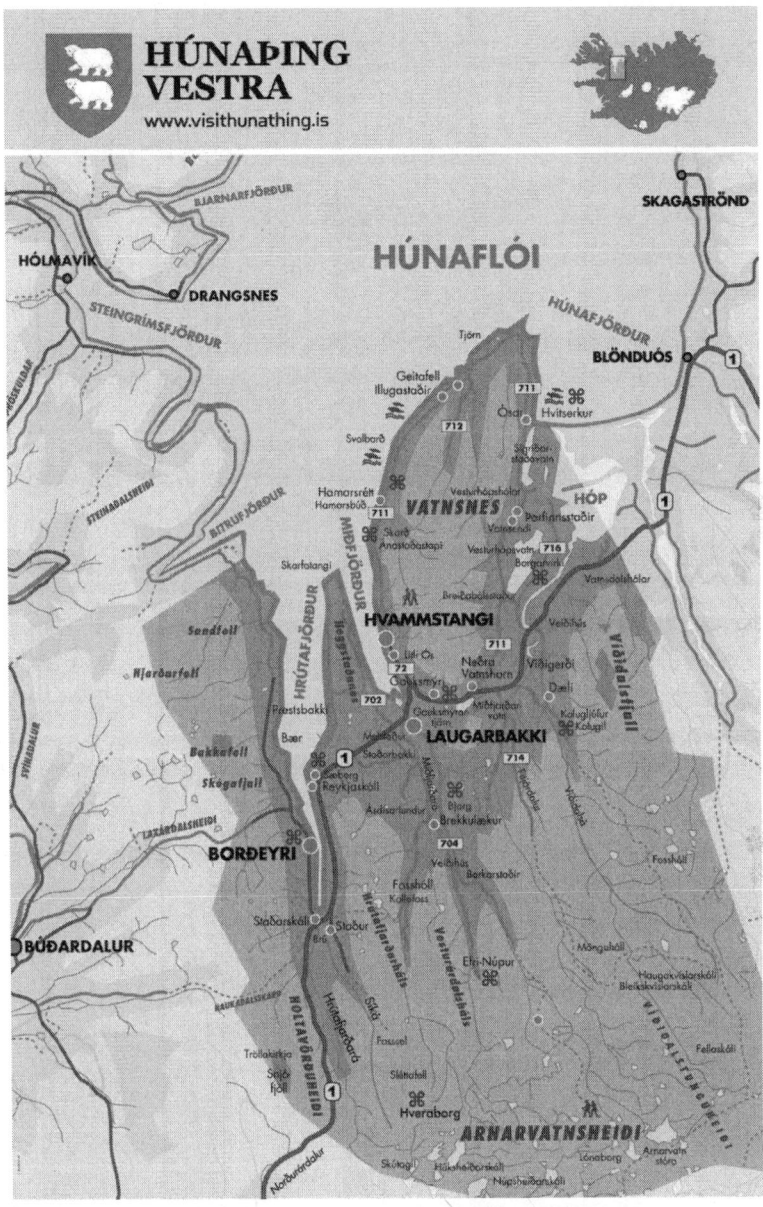

Source: http://www.visithunathing.is/is/hunathing-vestra/kort-af-hunathingi-vestra

who have gained citizenship and those who work in the municipality on a temporary, seasonal basis. After Hvammstangi, the next two most populated villages in Húnaþing vestra are Laugarbakki (population fifty-five) (Statistics Iceland 2019) and Borðeyri (around fifteen people; personal communication with resident, 4 February 2020). Approximately 50 per cent of the Húnaþing vestra population live on remote farms; in good weather, it may take up to forty-five minutes to travel from these locations to Hvammstangi (see figure 6.1).

The people of Húnaþing vestra, through particular community initiatives and engagement at both civil and local governmental levels, have worked at maintaining their cultural traditions and locally grown and produced products (Aquino and Kloes 2020). While there are many examples, one notable one is the Grettir the Strong project, which ran from 1999 to 2010. Aimed at preserving local history and cultural revitalization through the Saga of Grettir, the project successfully strengthened a sense of place and local identity. Community owned and operated, the project aimed to promote sustainable regional development and tourism as a tool for economic revitalization through community-based tourism. The project focused on using community assets for the preservation of local history and traditional storytelling and to sustain local pride. Similarly, the Icelandic Seal Center is a community initiative aimed at sustainable regional development, tourism as economic revitalization through community-based tourism, and seal research. The Icelandic Seal Center has used the strategy of an academic-community partnership for conducting research and projects. This approach recognizes the community as a social and cultural entity, and community partners are involved in all aspects of projects or research process (Johnson 2017). Through a partnership approach, the Icelandic Seal Center has helped the local community with the initiation of a preliminary seal-watching management plan at the grassroots level. Despite the lack of an official management plan from the Icelandic government, the community implemented a provisional code of conduct for seal watching on land and by boat, along with limited interpretive signs at seal-watching sites. This underlines how a partnership approach can co-create knowledge and change.

The Icelandic Seal Center serves as the gateway to seal tourism and research for visitors who want to learn more about seals. The centre has four emphases: a seal museum and gift shop (established in 2005), a research centre (established in 2007), a visitor information centre (established in 2012), and a travel agency (Seal Travel, established in 2016). The Icelandic Seal Center successfully lobbied the government in 2007 and 2008 to create two specialist positions for its research department: one to lead seal research in partnership with the Marine and Freshwater Research Institute, and the other to lead tourism research in partnership with Hólar University. In 2019, the research department added a naturalist position with a specialization in birds, in partnership with the Natural Institute of Northwest Iceland. The rationale was that bringing specialized people to the countryside could make a positive impact on local communities, build human capital, and also raise the profile of seal protection and sustainable wildlife tourism. This has come to fruition as the small population of Húnaþing vestra benefits from the specialists who now live in the area and have integrated with the community, establishing friendships and creating families.

Seal Travel was established by the Icelandic Seal Center to help strengthen a network of local tourism operators with a focus on nature, culture, and wildlife—and connecting these local operators to the regional, national, and international markets. Working closely with the community has the effect of creating active participants in authentic tourism experiences. After the creation of the Icelandic Seal Center, seals and seal-watching tourism was more widely recognized as an asset by the community, and the center continues to build the tourism industry in Húnaþing vestra after the Grettir the Strong project.

Size Matters?

The number of people in an area has a significant impact on a community's sustainability, resilience, and creative potential. For the purposes of this chapter, we conceptualize sustainability and resilience as separate but related terms in which sustainability is focused on mitigation and conservation and resilience entails a more adaptive and innovative approach (Burns 2018). Following Lew et al. (2018), we argue that both sustainability and resilience should be considered in development models for all communities.

For smaller communities, population numbers are critical to ensure sufficient quantity of resources can be pooled together and assets built to maintain a viable society. Larger communities—found in cities and urban areas—have more resources and more opportunities for full-time employment. The perception of life in these communities is often one of abundance. In contrast, living in a smaller town or in rural areas often has associated stigma linked to perceptions of poverty or lack of education and cultural refinement by others from larger areas. Smaller towns and rural areas, marked by lower population density, are also more vulnerable because of their remoteness from the cities or the capital, where much of the development policies are written. Additionally, they are more vulnerable to outmigration as young individuals and families move to larger cities for work or education. Consequently, smaller communities face greater challenges in terms of their capacity to both mitigate and adapt, and thus face a greater imperative to embrace creative strategies to cope with change.

In Iceland—as in many areas around the world—the perception persists that because larger settlements have more people, they have more qualified people, whereas smaller communities lack education/talent and skilled people. And yet, although many rural areas may have a lower percentage of highly educated people, this does not necessarily mean that rural people are unknowledgeable or untalented. Less tangible assets such as local knowledge of natural and cultural history, artistry, traditional handiwork and farming practices, and food preparation are important for community development as they add to the understanding and value of local resources. The art community is also filled with knowledgeable, creative, and talented people, and many are found in smaller remote areas of Iceland. For example, the Icelandic Textile Center in the town of Blönduós (population 867) has a well-known artist residency program. Seyðisfjörður (population 673), in the Eastfjords of Iceland, is known as the "arts village" and holds the yearly LungA Art Festival. Hvammstangi is home to Eldur í Húanþingi, an annual community and arts festival that attracts national visitors, and Handbendi Brúðuleikhús, an international puppet theatre company. In 2019, Leikflokkur Húnaþings vestra, Húnaþing vestra's theatre company, was chosen for a national award and performed at the National Theatre in Reykjavík (Ingilínardóttir 2019).

The stigma toward rural areas is a sign of a power imbalance vis-à-vis larger urban areas that ascribes a greater legitimacy to scientific knowledge over local knowledge. Smaller towns and rural areas are then further impacted because of political imbalances due to unequal access and power in management and decision-making (Chambers 2016). Exclusion from management and decision-making processes, even if unintentional, leads to tensions and mistrust, which further adds to political disengagement (Flannery, Healy, and Luna 2018). The Icelandic Seal Center, however, aims to address the need to manage human-wildlife interactions in tourism settings to ensure positive outcomes for all stakeholders—wildlife, local communities, and visitors. The centre works with both landowners and the local government to develop policies for the protection of seals and best practices for seal-watching tourism through local knowledge and scientific knowledge along with interdisciplinary and intersectoral research (see Aquino, Burns, and Granquist 2018; Burns et al. 2018; Granquist and Hauksson 2019; Granquist and Sigurjonsdottir 2014; Þorbjörnsson et al. 2017).

In Iceland, this "big versus small" tension is felt most strongly in the northern part of the country. Here, settlements are often smaller than those in the southern part of the country and farther away from Iceland's capital, Reykjavík, where the majority of the country's population lives and the majority of political decisions are made. For example, in 2018, Landsbankinn, an Icelandic financial institution with its main offices in Reykjavík, decided to shorten its hours and downsize their staff in smaller rural communities. Consequently, in addition to the shorter banking hours, the five full-time positions in Hvammstangi were reduced to two full-time and one part-time position. In the same year, VÍS, an Icelandic insurance company, closed all its offices in smaller rural communities in order to become more "digital." Businesses closing or scaling back hours have been met with heavy criticism because this disproportionality affects employment in rural areas (Einarsdóttir 2018). Rural municipalities argued that businesses that provide services for the entire country but locate their main offices in the capital region fail rural areas when they terminate the employment of people from the countryside (Húnaþing vestra 2018). To protest the closures, many municipalities are changing insurance companies (Kristjánsson 2018).

Two-thirds of the Icelandic population live in the Reykjavík Metropolitan Area, which has a population of 222,776 and a land area of 1,007 square kilometres. The total area of Iceland is about 103,000 square kilometres; thus, about 64 per cent of the population lives on 1 per cent of the total land area. The population of metropolitan Reykjavík is nearly twelve times larger than the second-largest settlement in Iceland, Akureyri, with a population of just 18,769[1] (Statistics Iceland 2019). Given Akureyri's size and urban status, it can be argued that it should be called a city. However, to many in Iceland, it is still considered a town (Iceland Magazine 2015), possibly because it is often compared with Reykjavík.

In Iceland, as in Europe more generally, the distinction between cities, towns, and villages varies widely. For example, distinguishing urban from rural areas is often based on the level of services available and the potential employment for people living in the surrounding areas, as well as population size (Konijnendijk et al. 2006). Size matters because the growth, or decline, of communities in general changes the mixture of residents, available housing, provision of services, and diversity and availability of jobs (Richards and Duif 2018). Urban areas are typically more densely populated, while rural areas tend to have lower population density (Konijnendijk 2003). In other words, if a settlement has community services, potential employment, and is more densely populated, then it earns the status of urban. Dennison and Ogilvie (2014) define cities as "settlement(s) with urban status and a population over 10,000. Small town = settlement(s) with urban status and population 2,000–10,000. Village/rural = settlement(s) lacking urban status and/or with population under 2,000" (656). Using these definitions, both Reykjavík and Akureyri would be classified as cities and the Municipality of Húnaþing vestra (with a population of approximately 1,200) would be rural. However, in Iceland, "settlements with more than 200 inhabitants are considered urban" (Benedikz and Skarphéðinsdóttir 1999 quoted in Konijnendijk et al. 2006, 98).

The concept of large versus small becomes important when residents perceive the value of their community in comparison with others. Drawing comparisons with other communities within their nation, and with cities in other nations, may affect their perceptions of the value of their own community, with the result that they might place a higher value on cities over rural areas. As the former mayor of Húnaþing vestra told the authors in March 2018,

> In Iceland, we are one of the smallest nations in the world. Okay, big country but very few people. We have the smallest capital. But still . . . sometimes [people in Reykjavík] do not trust small towns around the country for projects. And I can't understand why somebody living in the smallest city . . . does not trust anybody in a smaller town [for projects]. (Guðný Hrund Karlsdóttir, interview with authors, 14 March 2018)

Hvammstangi, as the administrative centre for the municipality, has a variety of services, including the state maternity office for the region, a playschool (ages one to five), a grade school (grades 1–10), and a distance upper secondary school, medical clinic and hospital, drug store, marketplace, and sports centre. Hvammstangi also contains a variety of businesses, including a restaurant, gas station and café, craft store, coffee shop, wool factory, museum, two mechanic shops, a world-class puppet theatre, and the Icelandic Seal Center. Other businesses are scattered throughout the municipality and along the Vatnsnes peninsula, where seal watching occurs. The maternity office is just one example that the former mayor of Húnaþing vestra, Gúðný Hrund Karlsdóttir, uses to demonstrate the capital area residents' lack of confidence in bringing government offices to the countryside: "I remember how the discussion was. A lot of people said that there was no way that you can have a good office here or anywhere in the countryside. Especially in Hvammstangi because we would never find qualified people. Silly!" (interview with authors, 14 March 2018).

Declining population in rural areas such as Húnaþing vestra, and northwest Iceland in general, has been a concern for some time. For example, while the total population of Iceland more than doubled from 71,000 people in 1890 to 194,000 in 1965, across the same years the rural population plummeted from 89 per cent of the country's total population to less than 18 per cent (Stone 1971). Factors attributed to this decline in rural areas, which had a major impact on the social landscape of Iceland, include the increasing importance of Reykjavík and Akureyri as business, administrative, and political centres, better transport routes, and new

Figure 6.2. Resident population trends in Hvammstangi from 1998 to 1 January 2019.

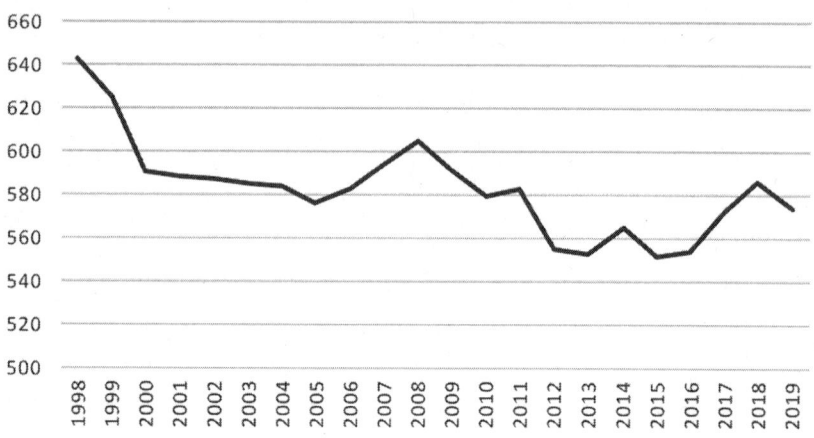

Source: Statistics Iceland 2019.

shipping technologies (Willson 2016). Since the early 1900s, commercial fishing moved from coastal farms to fishing communities (Kokorsch 2017), and by 1920 about 44 per cent of Icelanders lived in rural areas (Willson 2016). Although Hvammstangi has recently seen a revitalization of businesses and services, its population has yet to recover to the same level as 1998, when it had a population of 642 (see figure 6.2). Similarly, the population density of the Húnaþing vestra municipality has been maintained at around 1,200 since the year 2010 but has never reached the year 1998 population number of 1,412 (Statistics Iceland 2019). It is in this context, that of a small rural community grappling with depletion of population and services, that creative tourism has come to the fore.

Creativity as a Form of Resilience

Different definitions of creative tourism tend to vary in terms of their particular emphases; however, the concept usually includes "participative" and "authentic" experiences that allow tourists to develop their creative potential and skills through contact with local people and their culture (Richards 2011, 1237). A creative tourist can be seen as "the active co-creator or co-producer of their own experience" (Tan, Luh, and Kung 2014, 248). As an emerging creative tourism destination, Húnaþing vestra has harnessed tourism as a form of resilience, using it to create and build capacity for adapting to change and to events that negatively impacted the community and its members' livelihoods.

Burns (2018) describes Húnaþing vestra's capacity to adapt as mirroring that of wider Iceland, where, since settlement over a thousand years ago, communities have survived and thrived despite extreme weather events, frequent volcanic eruptions, plagues, and famine. However, more recent times have brought new and different challenges for Húnaþing vestra. For example, the introduction of fishing quotas across Iceland in the 1980s dramatically reduced the viability of Hvammstangi's fishing industry, along with other small fishing communities around the country. In the mid-1980s, sheep farmers suffered huge financial losses when most of their livestock were culled because of *riðuveiki* disease (known in English as "scrapie"). The introduction of dairy quotas in Iceland in 1980, and the subsequent increase in intensive farming practices, also had a lasting impact on the number of dairy farmers in the country, again inadvertently affecting smaller independent family farms. Between 1995 and 2007, the number of dairy farms fell by nearly half, while dairy production across the country more than doubled (Bjarnadottir and Kristofersson 2008). Dairy farmers able to stay in the business experienced an increase in their production because of growing demand caused by tourism (Arnarsdóttir 2015). The dairy factory in Hvammstangi closed in the late 1990s because of government restructuring, and the loss of jobs negatively impacted the community. As Gudrun Kloes, former tourism officer for Húnaþing vestra, explained, "We also lost a very good cheese. It was known all over the country, all over Iceland, 'Hvammstanga oustur.' It had a blue stamp on it [that said] 'Hvammstanga.' You could just walk in and buy it from the [dairy] factory" (interview with authors, 18 May 2018). Having a nationally recognized product made within your community can instil a sense of pride among locals.

Iceland's changing economy, coupled with the loss of some locally made products in Hvammstangi in the 1980s and '90s, may have impacted the perception of value, thereby contributing to a loss of community pride. Locally made products help reinforce a sense of place, build local identity and distinctiveness, and create a connection to the local community based on assets (Holtkamp et al. 2016); in turn, this builds pride (Aquino and Kloes 2020). Connection to place is important because it helps communities understand their assets and build a sense of pride, which, we argue, is linked to resiliency. Tourism more generally has used geography and a community's sense of place to connect visitors to the local identity via natural and cultural assets. Richards and Duif (2018), for example, describe placemaking as a "process of setting the available and potential resources of a place in motion by giving them meaning for the many actors who can use them to improve the quality of place" (17). In the following sections, we discuss how the community of Húnaþing vestra engaged with the many actors who saw the potential to nurture a sense of place using their natural and cultural assets to improve the quality of place. This led to creativity as a form of resilience in which tourism was used to enhance the local economy and community livelihoods by producing authentic and creative products for consumption within the tourism economy.

Tourism, Resilience, and Creativity in Húnaþing Vestra

Resilient communities are known for their creativity (Roberts and Townsend 2016), and creative people are often resilient. As the events in the 1980s and '90s, mentioned above, threatened the economic viability of small remote communities in Iceland like Húnaþing vestra, the country was simultaneously experiencing an increase in international visitors. Consequently, many rural families turned toward tourism as a coping strategy. This initially led to a boom in farm tourism (Jóhannesson and Huijbens 2010). Icelandic fishermen and farmers, already accustomed to supplementing their seasonal income (Arnarsdóttir 2015), found ways to harness tourism to their advantage. According to Hrafnhildur Yr Viglundsdóttir, the first managing director of the Icelandic Seal Center, "if you want to still live in the community and live on your farm you just have to find a way to do that" (interview with authors, 26 March 2018). For example, after suffering the loss of their sheep livestock because of scrapie, the farmers at Dæli in Húnaþing vestra needed to supplement their income. Out of necessity, and motivated by their interest in tourism, in 1988 they opened their farm as a guesthouse, which in addition to providing accommodation, included a restaurant, horseback riding facilities, and a tourist information centre. But the community did not stop at small-scale, individual-family-based farm tourism.

As outside visitors' interest in an area grows, increasing tourism provides rural residents with the opportunity to explore creative routes to enhance community resilience. Born in part from a desire, and perceived need, for competition, and associated with differential branding, tourism ventures sprung up across the Icelandic countryside with a distinctive, creative, and alternative flavour. Examples from northern Iceland include the Museum of Icelandic Sorcery and Witchcraft in Holmavik,[2] unique for being the "home of the necropants"; the Icelandic Sea Monster Museum in Bildudalur[3]; the Museum of Prophecies in Skagastrond[4]; and the Icelandic Seal Center in Hvammstangi,[5] which welcomes visitors to the "Land of Seals." All these locations have in common a fragile economy and potential social dislocation as a consequence of the economic downturn and outmigration. They are also the result of a shared public will to save their respective communities: a joint community vision to offer tourists something unique that will draw visitors to their area. Their success is due to the determination, and creativity, of the community members.

"Welcome to the Land of Seals"

Today, a sign in the shape of a seal at the turnoff to Hvammstangi via Route 1, which circumnavigates Iceland, welcomes travellers to the "Land of Seals," where several locations along the Vatnsnes peninsula allow access to harbour seal colonies. This includes one of the largest seal colonies in northwest Iceland, located near a rock formation called Hvítserkur, or "White Shirt," which has become a popular tourist attraction.

The Icelandic Seal Center was established in 2005 by a group of local residents who recognized a growing interest in seal-watching tourism but who also wanted to develop sustainable community tourism in Húnaþing vestra. However, the concept of the Icelandic Seal Center began brewing in the community a few years before its creation. According to Hrafnhildur Yr Viglundsdóttir, the first managing director of the Icelandic Seal Center, the area was by this point already gaining notoriety as the best seal watching location in Iceland. . . . People have been seal watching there since the 1960s and '70s. There had been a very active tourism officer there called Gudrun Kloes. . . . She is one of the idea makers of the Seal Center and one of the first people [who] worked with the idea with the people. And she started applying for funds for Svalbarð to make a walking path there. And then I think more and more people got into it . . . especially people from the Vatnsnes peninsula. (interview with authors, 26 March 2018)

The first public meetings with the wider community concerning the concept of the Seal Center began in April 2005. Gudrun Kloes indicated that it was difficult for the community at first to recognize their assets as a tourism destination: "They would just say, 'the north wind,' " essentially meaning that they could see no assets other than the cold north wind for which the area is known locally (interview with authors, 18 May 2018). However, Hrafnhildur Yr Viglundsdóttir explained that "there were some [assets] . . . and people connected with farming [and] connected with the peninsula . . . [and] people that were generally interested in seeing tourism grow in the area [began to realize that] this was something that was completely unique. Nobody else had focused on seals around Iceland. So, I think people saw a big opportunity to be something that could be a boom" (interview with authors, 26 March 2018). This marked the start of the community working to establish seal-watching sites along the peninsula and, correspondingly, encouraging other local businesses to engage with tourism. As the number of tourists began to rise, a number of businesses emerged to service the growing tourism market. The Icelandic Seal Center has had a positive effect on the community by helping to increase tourism to the area, which has also led to the revitalization of Hvammstangi, the Vatnsnes peninsula, and the Húnaþing vestra municipality.

Tourism in Húnaþing vestra and Hvammstangi focuses on seals, other wildlife (mainly whales and birds), and equine tourism, which are all easily recognizable in the region's place-marketing. A seal-watching boat, *Selasigling*, began operating on 25 February 2010 with a focus on seal and other wildlife watching (Sealwatching n.d.). Other businesses established along the peninsula include a restaurant, a café, and a handicraft market.

In 2012 the Seal Center also took on responsibility as the region's Tourism Information Center. Since 2013, accommodation built specifically for the tourism industry has been increasing; this includes the Hvammstangi Cottages (in 2013), Hotel Laugarbakki (in 2016), and the Hvammstangi Hostel (in 2017), adding to other farm stays, hotels, and Airbnbs within the municipality. The growth of many businesses has been a direct effect of the successful "Land of Seals" place branding.

Visitors to the Icelandic Seal Center are directed to designated seal-watching locations or other areas of regional interest. They can also visit the museum and learn how Icelanders have interacted with seals over time. The centre is a fifteen-minute drive from the closest seal-watching site and it aims to promote the development of tourism in the region through research and collaboration with tour operators and entrepreneurs; to promote the transfer of knowledge between the scientific community and society; to inform tourists about responsible seal-watching methods and viewing locations; and to develop environmental policy. The centre is managed and staffed by local residents and continues to co-operate with the community in its activities, which include an annual seal count and public lectures. It also co-operates with landowners and other businesses operating seal-watching sites. The Icelandic Seal Center has had a spill-over effect thanks to these collaborative efforts. Sharing information and research (locally and internationally), community capacity building (local lectures, workshops, and consultation), and connecting local tourism entrepreneurs to regional, national, and international markets (through Seal Travel) has helped to both diversify the local economy and build confidence in the tourism industry of Húnaþing vestra, nationally as well as internationally.

The growth in tourism businesses across Húnaþing vestra and the introduction of new and expanded services in Hvammstangi and Laugarbakki has led to a revitalization of the community. As shown in figure 6.2, resident population numbers in Hvammstangi were falling prior to the establishment of the Icelandic Seal Center. Shortly thereafter, these numbers rose slightly, before falling again, but they have now returned to almost the same as they were before the centre's establishment. This is apparent, for example, in the return of youth and families to the municipality because of the growing diversity of and opportunities for jobs. Young people who originally left the community to pursue higher education or better employment opportunities began to return and apply for jobs "that were suitable for their education" and experience (Gúðný Hrund Karlsdóttir, interview with authors, 14 March 2018). Young families have also returned, searching for a more rural setting in which to raise their children, with the possibility of paid employment in their field of training. Vercoe et al. (2014) refer to the process by which urban residents move into rural areas, often in search of idealized lifestyles, as "exurbanisation." In Húnaþing vestra, this pattern of returning community members, and the arrival of new ones, has helped to stem the tide of outmigration from the region.

Visitation to the Icelandic Seal Center increased from 4,958 visitors in 2011 to 31,019 visitors in 2019, with the majority of visitation in the summer months between May and September. Visitor numbers more than doubled between 2011 and 2012, the year the centre moved to a new building and included a Tourism Information Center. Matching both the growth and decline rate nationally, visitation to the Seal Center increased rapidly between 2015 (27,150 visitors) and 2017 (42,481 visitors) and decreased in 2018 (41,078 visitors) and 2019 (31,019 visitors) (see figure 6.3). On average, since 2012, about 35 per cent of the total number of visitors have paid to enter the museum. Entrance to the museum is free for members of the community and the centre hosts a number of free events for the community at the museum, furthering its relationship with locals.

The increase in visitation to the municipality since 2005 has both directly and indirectly affected services within the community. For example, during the summer season, the one general store in town is open every day with longer hours and a larger selection of items. Opportunities for restaurants and food services, especially during the summer season, throughout the municipality have also increased, with the first restaurant opening in Hvammstangi in 2015. These extra facilities benefit the locals as well as the tourists, as it "it makes it better to live in Hvammstangi to have these services. And we have two car repairs and we have the Brauðtofan [the bakery], the wool factory [Kidka], and we have Handbendi [the puppet theatre]. . . . We have services that would not be here if we didn't have tourism" (Gúðný Hrund Karlsdóttir, interview with authors, 14 March 2018).

Figure 6.3. Visitor numbers to the Icelandic Seal Center in Hvammstangi from 2011 to 2019.

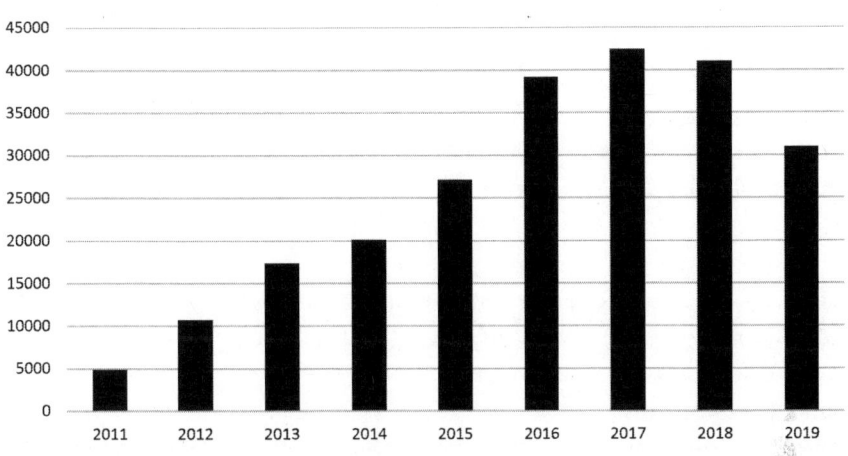

Source: The Icelandic Seal Center, reproduced with permission.

Creative Communities and Community-Based Tourism Development

Creative communities can be a powerful tool for addressing the social and economic needs of a society (Florida 2002a, 2002b; Foster 2009; Markusen and Schrock 2006; Zukin 2010). They can create, or recreate, a place's identity, thereby making it distinctive from other places, which in turn helps to carve out a niche in the marketplace (Richards 2011). Nurturing the arts and culture industries can help to revitalize both economic and social portfolios (Aquino, Phillips, and Sung 2012). For Húnaþing vestra, nurturing a sense of place identity and pride through seals as a form of natural capital helped to increase seal research and wildlife tourism management, entice job growth, develop better services for the community, create interconnections between wildlife tourism and business, attract skilled workers, and create spinoff and supporting businesses similar to arts-based economic development strategies. As Greta Clough, creator of Handbendi, a world-class puppet theatre based in Hvammstangi, put it,

> Creative community is for me . . . the place where everything comes from. The creative community goes beyond just the arts. It goes into everything. It goes into cultivating entrepreneurs, politicians, and . . . the influencers, because ultimately by investing in creative communities . . . you actually invest in the creation of imaginations and ideas and people who can think outside of the box and people who are interested in learning. (interview with authors, 20 March 2018)

The Icelandic Seal Center is an example of community-based tourism development driven by the efforts of community members—both in Hvammstangi and across Húnaþing vestra—to develop and manage their regional assets for tourism. The centre's board is made up of residents of Húnaþing vestra and regional stakeholders. The decision was made to include local community members to ensure that decisions were made in a community and regional context. The community-based tourism happening here exemplifies that which is community driven—effectively placing the decisions of what should happen or what is preferred in the hands of the community. The Icelandic Seal Center is consistent with the principles of sustainable development, but it goes beyond that to contribute to community resilience, and in that way, is also an approach for diversifying and revitalizing the local economy.

The members of the community who were interviewed for this chapter felt that the performing and visual arts, museums, academia, traditional handicrafts, storytelling, entrepreneurism, and leadership are all part of the creative industry. They also suggested that creativity is interconnected with community development, and that this creativity is the driving force behind the community's ability to create and produce experiences as a product. The role of creativity in the establishment and maintenance of tourism, and the associated ultimate goal of economic and social development for the region, was consciously and explicitly recognized. For the municipality of Húnaþing vestra, creativity was deliberately used to build a tourism product that could help support other seasonal income and to assist with the continuation of a way of life and the revitalization of a community. In the words of Hrafnhildur Yr Viglundsdóttir,

> Developing sustainable seal-watching tourism in the Vatnsnes peninsula was one of the main goals [of the Icelandic Seal Center], and contributing positively to the community of Húnaþing vestra as well as raising awareness about seals and the necessities of researching seals and protecting seals. . . . [W]e also had a clause about empowering the community, creating jobs, and making a positive impact on the economy. . . . Sustainability was always the first emphasis of the Seal Center, which, I thought, was very interesting that [the community] were aware of the importance of sustainable tourism development very early on. So, it was a guiding light for my work. (interview with authors, 26 March 2018)

The Icelandic Seal Center continues to bring both direct and indirect consequences for the community. For example, as Gúðný Hrund Karlsdóttir noted, young people coming to work in specialized jobs brought their partners/families, many of whom also have specialized skills and could further contribute to and shape the community. As tourism grew, it brought opportunities for the service industry to grow as well. Entrepreneurs fulfilling a need created new jobs. Tourism development in Húnaþing vestra highlights the interconnectedness of culture, creativity, and tourism innovation (Frey 2009; Richards 2011).

Since its beginning, the Icelandic Seal Center has actively engaged the community and helped to guide change by fostering public dialogue, to build community capacity and leadership, to contribute to the development of the community's knowledge of wildlife, and to encourage creative entrepreneurship. It has proven to be a powerful tool for community mobilization and activism. In many ways, the Icelandic Seal Center's goals to create healthy communities capable of action aligns with the argument for culture as an economic engine (Creative City Network of Canada 2005). The Icelandic Seal Center's success as a community-based tourism product lies in its capacity for engaging and involving the community from the beginning stages of tourism development planning through to the continuing maintenance of tourism products. In this case, the Icelandic Seal Center uses community-based tourism as a strategy for diversifying the local economy while also providing greater self-reliance and increasing local services.

Final Comments

For Húnaþing vestra, engaging with tourism as a form of community development led to the formation of an industry that involves the active participation of its residents as creators of authentic tourism experiences. Residents assumed a proactive role in building resilience through creative solutions, with the Icelandic Seal Center providing a model and path to attract highly skilled and educated people to live in the community and add to its assets, in the process diversifying its skills and further enhancing its creative potential. Establishment of this one centre fostered the uptake of other creative, co-creative, and locally produced products such as the puppetry theatre and businesses such as the seal-watching boat *Selasigling* and Seal Travel. Increased income from tourism has helped to create jobs in Húnaþing vestra, thereby assisting the reversal of problems associated with the rural-urban migration faced by many other small country municipalities in Iceland and around the world.

Húnaþing vestra's resilience is reflected in its community-based approach to developing seal-watching tourism as a product while also leading the way in seal research and protection. Creative tourism destinations offer the potential for enhancing the local economy and community livelihoods by producing authentic and creative products, and this is reflected in how Húnaþing vestra creatively integrated concepts from planning, sustainable development, and community development. It also reflects the fact that building collective capacity and providing tools for community engagement and activism is key to nurturing revitalization efforts. These efforts should go beyond focusing on economic impacts to also incorporate the need for community stakeholder involvement in order to shape the processes and outcomes of revitalization efforts. We argue that this is crucial as community-level redevelopment must have local engagement for longer-term sustainability. Mair and Reid (2010) conceptualize this as "an underlying assumption that long-term benefits of such an undertaking (sustainable community-based tourism) include not only a stronger community identity and the potential for community members to act proactively in the face of economic, social and cultural change, but a tourism product that is better grounded in community support" (408).

Aquino, Phillips, and Sung (2012) argue that having a sense of place and a strong identity is essential because this becomes the "calling card" for visitors. Beyond the scope of this chapter is a consideration of how residents from Húnaþing vestra identify with the "Land of Seals" branding or how much they feel that seals are a part of their culture. Further research is needed to understand Húnaþing vestra's community identity and the community's perspective toward the development, promotion, and conservation of cultural and natural heritage. This chapter has illustrated, however, the ability of locals to guide their desire for tourism development; it has also shown that the Icelandic Seal Center has encouraged the inclusion of stakeholders in community development processes and outcomes. Ensuring community engagement increases adaptive capacity building (Jurjonas and Seekamp 2018), enabling residents to achieve more of what they want for tourism development, as well as the resilience to address challenges.

Ultimately, this chapter provides insight into the fostering of community-based tourism built around the creative industry and resilience. It has demonstrated how talented residents can be proactive in building a creative tourism product and their strategies for continuing and creatively sustaining a rural community in northwest Iceland. The path chosen for and local support given to the Icelandic Seal Center provides a useful case study for other rural locations seeking to engage creatively, sustainably, and responsibly with tourism.

NOTES

1   This population estimate includes Akureyri and the surrounding area (Statistics Iceland 2019).

2   http://www.galdrasyning.is.

3   http://skrimsli.is.

4   http://www.sagatrail.is/en/museums/museum-of-prophecies.

5   http://selasetur.is/en/.

References

Aquino, Jessica F., Georgette L. Burns, Sandra M. and Granquist. 2018. "Seal Watching in Iceland: Ethical Management Development." In *The 9th International Conference on Monitoring and Management of Visitors in Recreational and Protected Areas (MMV9): Place, Recreation, and Local Development*, edited by Jeoffrey Dehez, 165–7. Bordeaux: MMV9. https://mmv9.sciencesconf.org/data/pages/last_version_abstract_book_7.pdf.

Aquino, Jessica F., and Gudrun M. H. Kloes. 2020. "Neolocalism, Revitalization, and Rural Tourism Development." In *Neolocalism and Tourism: Understanding A Global Movement*, edited by Linda J. Ingram, Susan L. Slocum, and Christina T. Cavaliere. Oxford: Goodfellow Publishers.

Aquino, Jessica, Rhonda Phillips, and Heekyung Sung. 2012. "Tourism, Culture, and the Creative Industries: Reviving Distressed Neighborhoods with Arts-Based Community Tourism." *Tourism, Culture and Communication* 12, no. 1: 5–18. https://doi.org/10.3727/109830412X13542041184658.

Arnarsdóttir, Eygló Svala. 2015. "Counting Sheep." *Iceland Review*, 20 August.

Bjarnadottir, Erna, and Dadi Mar Kristofersson. 2008. "The Cost of the Icelandic Transferable Dairy Quota System." *Icelandic Agricultural Sciences* 21: 29–37.

Blackstock, Kristy. 2005. "A Critical Look at Community Based Tourism." *Community Development Journal* 40, no. 1: 39.

Burns, Georgette Leah. 2018. "Searching for Resilience: Seal Watching Tourism as a Resource for Community Development in Iceland." In *Tourism Resilience and Adaptation to Environmental Change: Definitions and Frameworks*, edited by Alan A. Lew and Joseph Cheer, 51–67. London: Routledge.

Burns, Georgette Leah, Elin L. Öqvist, Anders Angerbjörn, and Sandra Granquist. 2018. "When the Wildlife You Watch Becomes the Food You Eat: Exploring Moral and Ethical Dilemmas When Consumptive and Non-Consumptive Tourism Merge." In *Animals, Food, and Tourism*, edited by Carol Kline, 22–35. New York: Routledge.

Cawley, Mary, and Desmond A. Gillmor. 2008. "Integrated Rural Tourism: Concepts and Practice." *Annals of Tourism Research* 35, no. 2: 316–37. https://doi.org/10.1016/j.annals.2007.07.011.

Chambers, Catherine. 2016. "Fisheries Management and Fisheries Livelihoods in Iceland." PhD diss., University of Alaska Fairbanks.

Chambers, Catherine, Guðrún Helgadóttir, and Courtney Carothers. 2017. " 'Little Kings': Community, Change and Conflict in Icelandic Fisheries." *Maritime Studies* 16, no. 1: 10. https://doi.org/10.1186/s40152-017-0064-6.

Creative City Network of Canada. 2005. "Making the Case for Culture: Culture as an Economic Engine." http://www.creativecity.ca/database/files/library/culture_economic_engine(2).pdf.

Dennison, Tracy, and Sheilagh Ogilvie. 2014. "Does the European Marriage Pattern Explain Economic Growth?" *Journal of Economic History* 74, no. 3: 651–93. https://doi.org/10.1017/S0022050714000564.

Einarsdóttir, Gréta S. 2018. "Insurance Company Closes Offices Around the Country." *Iceland Review*, 25 September. https://www.icelandreview.com/business/insurance-company-closes-offices-country/

Flannery, Wesley, Noel Healy, and Marcos Luna. 2018. "Exclusion and Non-Participation in Marine Spatial Planning." *Marine Policy* 88 (February): 32–40. https://doi.org/10.1016/j.marpol.2017.11.001.

Florida, Richard. 2002a. "Bohemia and Economic Geography." *Journal of Economic Geography* 2, no. 1: 55–71.

———. 2002b. *The Rise of the Creative Class: And How It's Transforming Work, Leisure, Community and Everyday Life*. New York: Basic Books.

Foster, Don. 2009. "The Value of the Arts and Creativity." *Cultural Trends* 18, no. 3: 257–61.

Frey, Oliver. 2009. "Creativity of Places as a Resource for Cultural Tourism." In *Enhancing the City: New Perspectives for Tourism and Leisure*, edited by Giovanni Maciocco and Silvia Serreli, 135–54. Dordrecht, NL: Springer.

Granquist, Sandra. M., and Erlingur Hauksson. 2019. *Population Estimate, Trends and Current Status of the Icelandic Harbour Seal* (Phoca vitulina) *Population in 2018* [Landselstalning 2018: Stofnstærðarmat, sveiflur og ástand stofns]. Reykjavík: Marine and Freshwater Research Institution, report number HV 2019-36.

Granquist, Sandra M., and Hrefna Sigurjonsdottir. 2014. "The Effect of Land Based Seal Watching Tourism on the Haul-Out Behaviour of Harbour Seals (*Phoca vitulina*) in Iceland." *Applied Animal Behaviour Science* 156: 85–93. https://doi.org/10.1016/j.applanim.2014.04.004.

Hall, Colin. 1991. *Introduction to Tourism in Australia: Impacts, Planning and Development*. Melbourne: Longman Cheshire.

Holtkamp, Chris, Thomas Shelton, Graham Daly, Colleen C. Hiner, and Ronald R. Hagelman III. 2016. "Assessing Neolocalism in Microbreweries." *Papers in Applied Geography* 2 (1): 66–78. https://doi.org/10.1080/23754931.2015.1114514.

Húnaþing vestra. 2018. "Byggðarráð ósátt við lokun útibús VÍS á Hvammstanga." Accessed 3 February 2020. https://www.hunathing.is/is/mannlif/frettir-og-auglysingar/tilkynningar-og-frettir/byggdarrad-osatt-vid-lokun-utibus-vis-a-hvammstanga.

Iceland Magazine. 2015. "Is Akureyri a Town or a City?" 31 August. http://icelandmag.is/article/akureyri-a-town-or-a-city.

Ingilínardóttir, K. D. 2019. "Leikflokkur Húnaþings vestra tekur yfir stóra svið Þjóðleikhússins." *Fréttablaðið*, 14 June. https://www.frettabladid.is/lifid/leikflokkur-hunathings-vestra-tekur-yfir-stora-svid-thjodleikhussins/.

Jóhannesson, Gunnar Thór, and Edward H. Huijbens. 2010. "Tourism in Times of Crisis: Exploring the Discourse of Tourism Development in Iceland." *Current Issues in Tourism* 13, no. 5: 419–34. https://doi.org/10.1080/13683500.2010.491897.

Jóhannesson, Gunnar Thór, and Katrín Anna Lund. 2017. "Creative Connections? Tourists, Entrepreneurs and Destination Dynamics." *Scandinavian Journal of Hospitality and Tourism* 2250 (June): 1–15. https://doi.org/10.1080/15022250.2017.1340549.

Johnson, L. R. 2017. *Community-Based Qualitative Research: Approaches for Education and the Social Sciences*. London: SAGE.

Jurjonas, Matthew, and Erin Seekamp. 2018. "Rural Coastal Community Resilience: Assessing a Framework in Eastern North Carolina." *Ocean and Coastal Management* 162: 137–50. https://doi.org/10.1016/j.ocecoaman.2017.10.010.

Karlsdóttir, Guðný Hrund. 2017. "Sveitarfélagið Mitt and Ferðaþjónustan." Hvammstangi. Presentation to the community.

Kokorsch, Matthias. 2017. "The Tides They Are a Changin': Resources, Regulation, and Resilience in an Icelandic Coastal Community." *Journal of Rural and Community Development* 12 (2/3): 59–73. Konijnendijk, Cecil. 2003. "A Decade of Urban Forestry in Europe." *Forest Policy and Economics* 5, no. 2: 173–86. https://doi.org/10.1016/S1389-9341(03)00023-6.

Konijnendijk, Cecil C., Robert M. Ricard, Andy Kenney, and Thomas B. Randrup. 2006. "Defining Urban Forestry—A Comparative Perspective of North America and Europe." *Urban Forestry and Urban Greening* 4, nos. 3–4: 93–103. https://doi.org/10.1016/j.ufug.2005.11.003.

Kristjánsson, Jón Þ. 2018. "Sveitarfélög hætta viðskiptum við VÍS." *RÚV*, 26 September. https://www.ruv.is/frett/sveitarfelog-haetta-vidskiptum-vid-vis.

Lew, Alan A., Chin-cheng Ni, Tsung-chiung Wu, and Pin T. Ng. 2018 "The Sustainable and Resilient Community: A New Paradigm for Community Development." In *Tourism Resilience and Adaptation to Environmental Change: Definitions and Frameworks*, edited by Alan. A. Lew and Joseph Cheer, 30–48. London: Routledge.

Mair, Heather, and Donald G. Reid. 2010. "Tourism and Community Development vs. Tourism for Community Development: Conceptualizing Planning as Power, Knowledge, and Control." *Leisure/Loisir* 31, no. 2: 403–25. https://doi.org/10.1080/14927713.2007.9651389.

Markusen, Ann, and Greg Schrock. 2006. "The Artistic Dividend: Urban Artistic Specialisation and Economic Development Implications." *Urban Studies* 43, no. 10: 1661–86.

Murphy, Peter. 1988. "Community Driven Tourism Planning." *Tourism Management* 9, no. 2: 96–104.

Reynarsson, Bjarni. 1999. "The Planning of Reykjavik, Iceland: Three Ideological Waves—A Historical Overview." *Planning Perspectives* 14, no. 1: 49–67. https://doi.org/10.1080/026654399364346.

Richards, Greg. 2011. "Creativity and Tourism: The State of the Art." *Annals of Tourism Research* 38, no. 4: 1225–53. https://doi.org/10.1016/j.annals.2011.07.008.

Richards, Greg, and Lian Duif. 2018. *Small Cities with Big Dreams: Creative Placemaking and Branding Strategies.* New York: Routledge.

Roberts, Elisabeth, and Leanne Townsend. 2016. "The Contribution of the Creative Economy to the Resilience of Rural Communities: Exploring Cultural and Digital Capital." *Sociologia Ruralis* 56 (2): 197–219. https://doi.org/10.1111/soru.12075.

Sæthórsdóttir, Anna Dóra, Colin Hall, and Jarkko Saarinen. 2011. "Making Wilderness: Tourism and the History of the Wilderness Idea in Iceland." *Polar Geography* 34, no. 4: 249–73. https://doi.org/10.1080/1088937X.2011.643928.

Sealwatching. n.d. "About Us." Accessed 19 October 2020. https://www.sealwatching.is/en/about-us/.

Simpson, Murray. 2009. "An Integrated Approach to Assess the Impacts of Tourism on Community Development and Sustainable Livelihoods." *Community Development Journal* 44, no. 2: 186.

Statistics Iceland. 2019. "Population by Municipality, Age and Sex 1998–2019: Division into Municipalities as of 1 January." Accessed 1 February 2020. www.statice.is.

Stern, Mark J., and Susan C. Seifert. 2010. "Cultural Clusters: The Implications of Cultural Assets Agglomeration for Neighborhood Revitalization." *Journal of Planning Education and Research* 29, no. 3: 262–79. https://doi.org/10.1177/0739456X09358555.

Stone, Kirk. 1971. "Isolations and Retreat of Settlement in Iceland." *Scottish Geographical Magazine* 87, no. 1: 3–13.

Tan, Siow Kian, Ding Bang Luh, and Shiann Far Kung. 2014. "A Taxonomy of Creative Tourists in Creative Tourism." *Tourism Management* 42: 248–59. https://doi.org/10.1016/j.tourman.2013.11.008.

Tao, Teresa C. H., and Geoffrey Wall. 2009. "Tourism as a Sustainable Livelihood Strategy." *Tourism Management* 30, no. 1: 90–8.

Vercoe, Richard, Meredith Welch-Devine, Dean Hardy, Jennifer Demoss, Shannon Bonney, K. Allen, Peter Brosius, D. Charles, Brian Crawford, S. Heisel, Heynen, R. G. de Jesús-Crespo, N. Nibbelink, Lamisha Parker, Catherine Pringle, A. Shaw, and Levi Van Sant. 2014. "Acknowledging Trade-offs and Understanding Complexity: Exurbanization Issues in Macon County, North Carolina." *Ecology and Society* 19, no. 1: 23. https://doi.org/10.5751/ES-05970-190123.

Willson, Margaret. 2016. *Survival on the Edge: Seawomen of Iceland.* Seattle: University of Washington Press.

Zukin, Sharon. 2010. *Naked City: The Death and Life of Authentic Urban Places.* New York: Oxford University Press.

Þorbjörnsson, Jóhann. G., Erlingur Hauksson, Guðjón M. Sigurðsson, and Sandra M. Granquist. 2017. *Aerial Census of the Icelandic Harbour Seal (Phoca vitulina) Population in 2016: Population Estimate, Trends and Current Status.* Reykjavík: Marine and Freshwater Research Institute.

Placemaking through Food: Co-creating the Tourist Experience

Susan L. Slocum

Introduction

Culture is increasingly being used to enhance tourism destinations and create economic opportunities through new attractions, cultural routes, and heritage centres. Urry (2001) argues that culture has become an essential element of tourism product development, which has made cultural tourism one of the fastest-growing segments of the leisure tourism market (Richards 2009). This growth is attributed to the high spending patterns and longer travel durations associated with cultural tourists, who often spend one-third more than other types of travellers (Richards 2009). However, the rise in cultural and heritage tourism products has resulted in the reproduction of formulaic cultural narratives as destinations compete with each other to fulfill tourist expectations of knowledge acquisition (Richards and Wilson 2006). The demand for cultural immersion, rather than cultural sites, has given rise to the creative economy in an attempt to meet the needs of a changing consumer travel base.

Creative tourism is a form of cultural tourism in which identity forms the basis of both production and consumption. From a supply-side approach, creative tourism is the provision of memorable experiences in a way that portrays an authentic narrative of a lived experience (Binkhorst and Den Dekker 2009). While recognizing that "authentic" is a much-contested concept, tourism producers try to accurately recount their culture in a way that allows consumers to experience their reality and identity.

From the demand side, Richards (2011) defines creative tourism as a participatory experience that develops a tourist's creative possibilities through contact with local people and their cultures. Creative tourism is thus the "truth learned from personal experience with a phenomenon rather than truth acquired by discursive reasoning, observation, or reflection on information provided by others" (Borkman 1976, 446). Creative tourism is the communication of culture that influences the identity of both the tourism provider and the visitor.

Food tourism is often perceived as a natural outgrowth of creative tourism. However, past literature has presented either a supply-side approach or demand-side analysis, and has neglected that the experience of food and drink consumption is a shared practice. Through the lens of sustainable tourism, it is the co-construction of consumption that ultimately brings success to small communities, described as granular communities by Scherf in this book's introduction, using food tourism as a development tool. Therefore, this chapter calls for a broader approach to food tourism studies as a means to reflect on the shared construction of food experiences, and it situates that approach as a potential method for creative tourism development and agent of change by including all the themes embedded in this book: (1) co-creation by visitor and resident of experiences that feature unique local skills and knowledge; (2) engagement of visitor imagination by participation in tangible or intangible endogenous culture; (3) generation or regeneration of sustainable cultural development for the host community; (4) formation of creative networks to offer touristic experiences; (5) examination of the processes, policies, and methodologies around creative tourism; and (6) creative representation of smaller communities. Using a literature review format, the goal is to combine supply and demand attributes from a variety of studies to show the underlying benefits of assessing the co-creation of food tourism through product partnerships. This chapter argues for the inclusion of both producers and consumers as a means to enhance the potential benefits that can arise for small communities embarking on food-and-drink tourism.

Food Tourism as Sustainable Tourism

Global food policy has had a dramatic effect on rural areas. Export strategies have left small-scale agriculture increasingly dependent on the international food industry as a diversification strategy, resulting in unequal power distributions, higher entry barriers, and the distribution of financial rewards toward developed nations (Bolwig et al. 2010). Increasing production costs and decreasing prices have led to the reduction of family farms on a global scale, changing the pastoral landscapes and impairing the economic opportunities of rural regions (Ilbery et al. 2005). Monoculture farming has diminished traditional agricultural practices and created food deserts in areas that traditionally grew a variety of food items to support the needs of neighbouring urban communities. Moreover, monoculture requires the extensive use of fertilizers and pest-control measures, which erode the environmental health of these communities (Slocum and Curtis 2017b). As rural regions face increased hardships, such as outward migration, aging populations, and a lack of investment in infrastructure, heath care, and education, food tourism is recognized as a viable development option, incorporating open space opportunities with food traditions as the basis for the tourism product. Innovative tourism markets have allowed small communities new opportunities to expand traditional agricultural production and increase diversity as a means to mitigate these economic challenges.

These production changes have also led to changes in consumer demand. The lack of transparent oversight in global food chains has increased fears around food safety and quality, specifically the use of chemicals in monoculture farming and ripening agents for long-haul transport (Henson and Humphreys 2010). Research has shown that increased freshness of produce, better taste, and a perceived healthier alternative is a primary driver in local food consumption (Delind 2006; Schnell 2011). Sustainable consumption includes the use of products that bring a better quality of life while minimizing the use of natural resources and toxic materials, as well as reducing greenhouse gas emissions, waste, and pollutants (Seyfang 2006). There is evidence that food consumers in developed countries are motivated by intrinsic (food quality and appearance), extrinsic (retail and shopping experience), and credence (healthiness, environment, and rural welfare) elements, each of which have spurred the "buy local" movement (Weatherell, Tregear, and Allinson 2003).

Early in the discussion of food tourism theory, Hall et al. (2003) recognized the sustainable development potential of food tourism. The incorporation of local food in the tourism sector has been acknowledged as a form of sustainable tourism because it strengthens local production through backward linkages in the tourism supply chain, resulting in decreased economic leakages and socio-cultural impacts (Slocum 2015). Slocum (2015) writes, "Eating habits can be viewed as an expression of culture and the preservation of traditional foodways is a substantial part of cultural identity" (245). Tourists seek new food-based traditions as a way to connect to the cultures they are visiting (Everett and Aitchison 2008), which has the potential to instill pride in local populations vis-à-vis their traditional agricultural practices and food traditions. Moreover, the growth in organic produce and the desire to eat food close to the area in which it is produced has reduced the environmental impact of food production. It is not surprising that food and drink have become a foundation of tourism marketing (Schnell 2011).

The Creative Economy

Howkins (2002) describes the creative economy as a system in which added value is instilled through imaginative qualities, rather than through the use of the traditional resources of land, labour, and capital. This concept initially referred to the arts, media, games, and research and development, where creative input is used as the primary source of value and the cause for economic transactions. Richards (2011) highlights the natural symmetry between the creative economy and tourism when he writes that

> The practice of tourism . . . involves all four [creative] approaches, for example in the use of the creative environment through visits to creative clusters, the use of creative products as tourism attractions [e.g., travel related to famous authors, painters, etc.], the utilization of the creative process in designing creative activities for tourists (e.g. workshops and masterclasses) and the involvement of creative people through the activities of the "creative class." (1226)

In recognizing that tourism is experiential in nature, and that it is produced and consumed simultaneously (Pantzar and Shove 2005), tourism has become a driver of the creative economy (Richards 2011). As tourism continues to rely on staged experiences for mass consumption, the tourism product and the destination are requiring new forms of innovation as a means of differentiation.

Tourism is dependent on a region's natural and cultural resources. These resources create a sense of place, or the special meanings that represent the identity and character of a region (Lockie 2001). Sense of place incorporates the physical setting, human activities, and human social and psychological processes within the natural environment that describe our attachment to and dependence on a place and its identity. Sense of place involves human experiences, social relationships, emotions, and thoughts that represent a communal identity (Liu and Cheung 2016). It is this identity that many tourists seek when travelling. Thus, it is the communication of sense of place that forms the basis of the experiential tourism product and the tourism narrative in food tourism. Creative tourism is essential in telling the story of the subtle differences in a destination's sense of place and communicating local identity in a way that can be consumed by tourists.

Yet, sense of place is not static, and tourist interactions with communities have the potential to change a destination's essence. In urban contexts, creative industries are being used to "transform" living spaces into areas for play. Marques and Borba (2017) acknowledge that "The interpretation and practices of a 'playable city' relate to co-creation in the way the city is made playable [bottom-up projects] but also in the way it is experienced, both by residents and tourists, as it promotes the interaction with the space" (87). In the same light, creative industries can change rural areas, redefining the role of open space and idyllic living through tourism consumption. There are opportunities for socio-environmental change through deeper understandings of consciousness and reconnection of the culture-nature divide through food tourism production (Cavaliere 2017). The creative formulation of the tourism narrative can shape traditional ways of knowing, creating new realities and expunging outdated notions of place identity. Serving tourists and creative talent development may present changing identities for communities accustomed to traditional industries (Blapp and Mitas 2017).

Food Tourism as Creative Production

Food is food. The ingredients in most recipes are similar across regions and cultures, starting with the basic grains (rice, wheat, corn, barley), combined with vegetables (tomatoes, potatoes, cucumbers, peppers), a few spices (oregano, coriander, cumin, garlic), and perhaps a helping of meat (beef, chicken, pork) or fish. There are exotic foods that may be unique to an area (e.g., sea cucumber, squid-ink pasta, haggis), but it is the combination of common ingredients in a variety of ways that constitute most of the food eaten around the world. However, the concept of foodways helps to describe the subtle distinction between eating for nutrition and eating for experience. Foodways suggest that food is a network of activities: the production of raw materials; the planning of a menu; the preparation of the ingredients; the presentation and performance of eating; the preservation or repacking of excess items; and the reflection or discussion about past meals (Long 2004). It is within a region's foodways that distinctiveness arises, as Long (2004) aptly expresses: "Since food is more than the dishes we eat, we can be tourists by exploring these other aspects of the food system. It means we can mix the new with the old, the exotic with the familiar" (23). Foodways provide the channel to communicate sense of place.

By viewing food as a system, it is easy to see the creative opportunities inherent in food preparation. Food has become an artisan product, one that is produced with pride and that showcases a region's complex sense of place. Innovation can come through the growth of new or traditional varietals (e.g., heirloom), through cooking techniques (e.g., fusion cooking), or through food-based experiences (farm visits or cooking classes). Food is derived from the natural environment and provides insight into the histories, ethos, and identities of the cultures within that environment, thereby enhancing the sustainability of tourism (Long 2004). But the concept of foodways is not only about eating food; it also allows for experiences such as volunteering on a farm, shopping for handcrafted culinary souvenirs, attending a craft beer festival, or touring a food production facility.

Slocum and Curtis (2017a) highlight the entrepreneurial talents of farm shops in England. As brokers and promoters of local food, farm shop managers have the flexibility to negotiate a highly competitive and changing business environment. They write, "[they] quickly adjust the focus of their business, experiment with new product and service offerings, and adjust their business model to accommodate changes according to the needs of their clientele" (47). In order to keep their shelves stocked, they collaborate with growers to find innovative products, encourage new entrants to ensure variety, use branding strategies to define "local," and determine the local food image presented to consumers. As lifestyle businesses, they collaborate with festivals, farmers' markets, and other outlets to support enhanced food-related experiences for residents and visitors alike. Their creative input vis-à-vis retail operations constructs the narrative of local food and their partnership requirements enhance social capital around food production.

Another creative example is presented by Azizi and Mostafanezhad (2014) by means of on-farm volunteer opportunities for tourists in Hawai'i. Through WWOOF (World Wide Opportunities on Organic Farms), volunteers live and work at organic farms specifically to learn about organic agriculture. Through a "social movement approach" (148), farmers are able to use tourism labour, specifically young people looking for alternative ways of life, to provide healthy, affordable, and environmentally friendly foods to local residents as well as restaurants, markets, and stores that support tourism growth. Through a strict screening process that seeks participants with "a genuine and honest interest in people" (147), hosts understand the tourist's background, their expectations for spiritual and cultural growth, and design hands-on experiences in growing and marketing handcrafted food products. Moreover, "farm volunteering can provide a direct pathway for tourists to the backstage of a community . . . by avoiding the cash nexus outside the commodified relations of the tourism industry" (136). Volunteers help reduce the need for business expansion, maintaining the lifestyle business and developing the creative talents of the farm and their guests.

Food tourism has the potential to foster the "rediscovery" of food traditions, as emphasized by Grasseni (2014) in the northern Italian Alps. The use of geographic indicators can often lead to standardized products that are viewed by locals as lower quality. By drawing attention to the "rebel" (57) cheese makers in Valli del Bitto, Grasseni acknowledges that diversity is an "endangered resource" (59) unless a region can find an economically sustainable path to cultural resilience, as highlighted by de la Barre in chapter 4 and Aquino and Burns in chapter 6 of this volume. The "ecomuseum" designation made by the local government has "become what common sense would have it: an open air park, complete with three routes, [that] provides an ecological packaging of cultural and environmental diversity for advertising the valley's produce" (59). When tradition is transformed into a method of diversity, creative tourism opportunities arise through the combination of immaterial resources and material culture in which traditions are reinvented as a means to add value to local products. The resurgence of alternative forms of production that differentiate the product and align with the superior quality values of the community can be developed.

These three examples provide a general overview of how smaller communities are using the creative economy within the context of food. Each provides opportunities for memorable experiences, while showcasing regional identity in creative and artistic ways (Binkhorst and Den Dekker 2009). As is common in rural agricultural regions, these businesses are lifestyle endeavours (Slocum 2015), and tourism has provided a way for small farms to ensure adequate profit without the requirement of extensive growth or capital investment (Howkins 2002). Not only do these examples emphasize creative products distributed in creative environments, they show the creative process through which creative people construct and transmit their own evolving identities (Richards 2011).

Food Tourism as Creative Consumption

Tourists must eat. That said, what tourists choose to eat varies from individual to individual. Many studies have shown trends in tourists' eating preferences, attempting to group similar tourists together as a means to delineate the food tourism market. For example, one early study grouped travellers into two groups, neophobes, or people who fear anything new, and neophiles, ones who like to experiment and try new things (Pliner and Hobden 1992). Shenoy (2005) applied these concepts to food tourists, acknowledging that tourist may fall somewhere on a spectrum between neophobe and neophile. Food preferences can vary based on the personal characteristics, activities, and motivations of tourists, and there is evidence that food choices are inspired by visitor nationality, the destination visited, seasonality, cultural experiences, interpersonal relations, sensory appeal, and health concerns (Slocum and Curtis 2017b). However, as Slocum and Curtis write, "Outside of specific food-related travel, there is very little data to provide more than generalizations about tourists and their food-related habits" (129). Moreover, psychographic characterizations are less common than demographic or geographic analyses, leaving many unanswered questions about the motivation for food-related travel.

Food tourists are defined as people who travel to destinations that have established reputations as places to experiment with quality local food (UNWTO 2012). Some food tourists may treat food as a small part of the whole travel experience, while others may select destinations based on their food traditions and reputation (Hall et al. 2003). Research has shown that food tourists are generally open to new experiences, desire life-long learning, and are educated consumers (Croce and Perri 2010). They also value cultural immersion and participate in outdoor actives. There is evidence that food tourists care about sustainability, although there is no understanding of the cause-and-effect relationship between food preferences and sustainability values. For example, do sustainably minded individuals use their food choices to promote their values, or is it exposure to new foods (and culture) through travel that encourages pro-sustainability ideals?

In a study of long-term backpacking tourists, Falconer (2013) shows how the narratives of long-term change are embodied during the journey through food. She highlights that food provides the "structure and rhythm" (34) to life on the road, where the body wavers between the novelty of new food experiences that become routine over time, leaving the traveller wanting comfort foods from home. As these longings are fulfilled, the traveller weaves back to the exotic with a replenished sense of experimentation. Falconer writes, "tourists make sense of spaces through the sensual dimensions of food and drink: through the nostalgia of taste and smell, romanticised imaginaries of 'exotic' or 'homecomfort,' or the contrast between 'local' and 'global' foods" (34). As travellers negotiate their global cosmopolitanism alongside their cravings, their identity and expectations are shaped by the experience.

Bessiere and Tibere (2013) show the importance of food in the development of identity for tourists in France. Local recipes are seen as "symbolic consumption" (3425) of a region, its land, and its people. However, it is the hospitality of the personal encounters that forms the connection to place. Hospitality provides the conversations that transmit the narratives of place and identity. These connections occur in restaurants, markets, on farms, and in wineries. The authors note that many regions use branding strategies to "enhance the attractiveness of traditional foodstuffs" (3424), and that souvenirs are purchased in places where heritage contact is perceived as the most impactful. The authors write, "Initiation into food cultures thus results from a dialogue with the natural environment providing a code to be deciphered by tourists as they make their way into the region" (3425). Some tourists describe a journey of self-discovery, while others connect to a country or region, emphasizing "the potential role of [food] tourism in the process of change" (3425).

Rudy (2004) provides another interesting study related to Mormon missionaries during their assignments abroad. Part of the goal of missionary travel is to engage in prolonged interactions in an unfamiliar culture to establish a connection that may bring new members to the church. Often, food is the vehicle that links people with their host culture, making food indistinguishable from the relationships they form while travelling. In turn, the sharing of food can reward the traveller "with personal growth, empathy, and long-term, life-changing experience in creating a community and comprehending the unfamiliar" (154). Rudy claims that the process of turning unfamiliar foods into familiar ones provides an opportunity to extend interpersonal understandings and expand personal tastes.

These three demand-side studies help to illuminate the process of change and personal growth that food tourism can provide for tourists. Travellers recognize the value inherent in food through the relationships derived around the food preparation and eating process (Howkins 2002). Within the consumption of a region's identity and culture, tourists are establishing their own sense of place through the context of food and food practices (Lockie 2001). Liu and Cheung (2016) emphasize that sense of place involves human experiences, social relationships, emotions, and thoughts as a way for travellers to connect with a communal identity, one that now includes a changed self as the tourists negotiates her own self-construction. It is the participatory nature of food tourism that is simultaneously being produced and consumed that situates food tourism squarely within the realm of the creative economy.

Discussion

Creative Food Tourism and Sustainable Rural Development

There is ample evidence of the sustainability benefits of food tourism. Food tourism has been shown to mediate economic challenges in rural areas through backward linkages in the tourism value chain, reducing poverty by providing jobs and opportunities for small businesses (Hall et al. 2003). It also has the potential to reconstruct social narratives that reinvent traditions and celebrate culture (Long 2004). Moreover, consuming food as close as possible to production sites reduces greenhouse gas emissions and the need for added chemicals to maintain freshness during transport (Henson and Humphreys 2010). Food tourism has the potential to support "lifestyle characteristics that unite environmental stewardship with support for local economies and emphasizes a 'slow' approach and sense of place within the tourism experience" (Slocum 2015, 252). However, it must be noted that the sustainability of food tourism is dependent on the practices of local growers and food-service businesses (Slocum 2015).

Moreover, the food tourism narrative has the potential to increase awareness of social and environmental conditions (Cavaliere 2017). Food tourists are educated travellers seeking an exchange of knowledge through the consumption of cultural experiences (Croce and Perri 2010). They value sustainability goals, participate in outdoor recreational activities, and are willing to spend more money on responsible tourism products (Slocum and Curtis 2017b). They derive value from exploring foodways and seeking a cultural product that is sensory in nature and that transmits the sense of place and the lived experiences of the communities they visit (Howkins 2002).

Creative Food Tourism as an Agent of Change

The case studies presented in this paper support the theory that food tourism lies within the realm of the creative economy, communicating economic and socio-cultural values through creative production and consumption. Using a crafted narrative, food tourism creates memorable experiences that have the potential to influence a traveller's sense of self (Binkhorst and Den Dekker 2009). Moreover, food tourism incorporates Richards's (2011) four creative elements: creative environments, creative products, creative process, and creative people. Rather than rely on the traditional resources of land, labour, and capital, food tourism uses creativity as the added value, creating impactful relationships as the basis for the tourism product (Howkins 2002).

Yet, this chapter argues that the current literature fails to measure the co-constructed narratives that appear to dominate the food tourism experience. The primary flaw in the supply-side cases is that the voice of the tourist is silent, with the result that we as scholars do not know the effectiveness of the narratives being described or if the "truth" is being learned from personal experiences (Borkman 1976). The demand-side cases show spiritual changes in the tourists themselves, but they neglect to show exactly which creative elements in food tourism production influences this change. One could argue that the simple action of communal eating brings on these transformations, yet these experiences are "crafted" with tourists in mind, and no two food tourism experiences are identical. Food tourists seek cultural immersion that includes knowledge transfer, which implies that the simple process of eating strange or unfamiliar foods is not enough to provide a conduit for personal growth. This chapter proposes that it is the narrative associated with the experience that transmits a sense of place (Binkhorst and Den Dekker 2009), and the narrative evolves as actors interact and change throughout the creative process (Marques and Borba 2017). This narrative, while crafted by the tourism provider, is nevertheless interpreted and realized by the tourist.

Often, the values possessed by the tourism provider is the primary motivation for change within food tourism product development. For example, the organic farms in Hawai'i realized that their tourism product required a very specific type of visitor, one that valued the learning experience and who would contribute to the social movement approach inherent in their business model (Azizi and Mostafanezhad 2014). This implies that not all tourists are open to receiving the sense-of-place narrative valued by the creative entrepreneurs. Slocum and Curtis (2017a) provide another example in their case study wherein a farm shop owner ended a contract with a local garlic producer because he was purchasing his garlic from overseas, even though he was crafting his product locally. By recruiting a new entrant into the industry, the farm shop was able to supply locally sourced *and* locally crafted garlic products. Therefore, rather than selling garlic items, the farm shop is defining local food through the narrative that expresses their commitment to local agriculture (Long 2004). In both cases, the tourism product is deeply rooted in sustainability ideals that inform supply decisions. However, the impact on demand decisions is not addressed in these cases.

The social interactions that are formed through the production and consumption of food tourism have the potential to provide open channels of communication in an intimate setting. Liu and Cheung (2016) emphasize the importance of social relationships in identity formulation; they argue that the mere presence of tourists has the potential to influence identity and sense of place. The co-creative nature of food tourism allows for quick adjustments to the experience in real time by providing an instantaneous feedback loop, potentially changing the food tourism narrative along the way. Marques and Borba (2017) write, "The co-creative processes should continue at different levels in the interaction with residents and tourists, improving their relationship to the [region] and changing their perspectives" (90). What constitutes "good food" is inherently different from what constitutes a "good food experience." Further discussions on how food tourism influences sense of place is therefore warranted.

Richards and Wilson (2006) acknowledge that change is a fundamental aspect of the creative economy, which happens within both the production and consumption of food tourism. Yet, the nature of change is inherently different. Both Bessiere and Tibere (2013) and Rudy (2004) highlight changes in visitors that involve self-identity and discovery, yet the change experienced by suppliers appears to be related to their business models rather than perceived at the personal level. As lifestyle businesses, these creative entrepreneurs are seeking more than just profits; indeed, quality of life is an essential motivator (Azizi and Mostafanezhad 2014; Slocum and Curtis 2017a). Intuitively, the social bonds formed through food tourism interactions must have some impact on creative entrepreneurs at a personal level (Cavaliere 2017), but again, this voice is silent in the literature. Moreover, the change in sense of place and how that is manifested in food tourism destinations remains obscure in the current literature (Blapp and Mitas 2017; Marques and Borba 2017).

Conclusion

Food tourism has become a major contributor to tourism studies, especially among scholars focusing on rural areas where food traditions are entwined with the history, heritage, and identity of communities. Collaborative networks, both among producers and between producers and consumers, have given rise to a new culture of creativity in small communities. For areas that may not have traditional tourism assets (national parks, large-scale attractions, or all-inclusive resorts), these case studies highlight how food has united agricultural communities by instilling pride through the communication of a sense of place. Similar to Prince, Petridou, and Ioannides in chapter 10 of this volume, this chapter has specifically argued that for small rural communities, the crafting of narratives has the potential to provide an inclusive self-determination that allows for local control over tourism development.

As scholars, our job is to create new knowledge around social relationships. While understanding visitor characteristics and motivations is important for the development of the tourism industry, recognizing how tourism and host interactions change the nature of a rural region is also vital to enhancing quality of life and social sustainability. The very nature of tourism impacts, such as increased traffic, changing food practices, and new open space usage, can have profound effects on a region's sense of place and foodscapes. Identifying how this change is negotiated through the creative economy and the tourism narrative can provide insight into future rural development challenges. This chapter is a call to action aimed at enhancing the potential benefits that can arise for small communities embarking on food-and-drink tourism. ·

References

Azizi, Saleh, and Mary Mostafanezhad. 2014. "The Phenomenology of World Wide Opportunities on Organic Farms (WWOOF) in Hawai'i: Farm Host Perspectives." In *Rural Tourism: An International Perspective*, edited by Kathrine Dashper, 134–50. Newcastle upon Tyne, UK: Cambridge Scholars Publishing.

Bessière, Jacinthe, and Laurence Tibere. 2013. "Traditional Food and Tourism: French Tourist Experience and Food Heritage in Rural Spaces." *Journal of the Science of Food and Agriculture* 93, no. 14: 3420–5.

Blapp, Manuela, and Ondrej Mitas. 2017. "Creative Tourism in Balinese Rural Communities." *Current Issues in Tourism* 21, no. 2: 1–27. https://doi.org/10.1080/13 683500.2017.1358701.

Bolwig, Simon, Stefano Ponte, Andries du Toit, Lone Riisgaard, and Niels Halberg. 2010. "Integrating Poverty and Environmental Concerns into Value-Chain Analysis: A Conceptual Framework." *Development Policy Review* 28, no. 2: 173–94.

Borkman, Thomasina. 1976. "Experiential Knowledge: A New Concept for the Analysis of Self-Help Groups." *Social Service Review* 50, no. 3: 445–56.

Binkhorst, Esther, and Teun Den Dekker. 2009. "Agenda for Co-Creation Tourism Experience Research." *Journal of Hospitality Marketing & Management* 18, nos. 2–3: 311–27.

Cavaliere, Christina. 2017. "Foodscapes as Alternate Ways of Knowing: Advancing Sustainability and Climate Consciousness through Tactile Space." In *Linking Urban and Rural Tourism: Strategies in Sustainability*, edited by Susan L. Slocum and Carol Kline, 49–66. Wallingford, UK: CABI.

Croce, Erica, and Giovanni Perri. 2010. *Food and Wine Tourism: Integrating Food, Travel and Territory*. Wallingford, UK: CABI.

Delind, Laura. 2006. "Of Bodies, Places, and Culture: Re-situating Local Food." *Journal of Agricultural and Environmental Ethics* 19, no. 2: 121–46.

Everett, Sally, and Cara Aitchison. 2008. "The Role of Food Tourism in Sustaining Regional Identity: A Case Study of Cornwall, South West England." *Journal of Sustainable Tourism* 16, no. 2: 150–67.

Falconer, Emily. 2013. "Transformations of the Backpacking Food Tourist: Emotions and Conflicts." *Tourist Studies* 13, no. 1: 21–35.

Grasseni, C. 2014. "Of Cheese and Ecomuseums: Food as Cultural Heritage in the Northern Italian Alps." In Giovine *Edible Identities: Food as Cultural Heritage*, edited by Ronda L. Brulotte and Michael A. Di Giovine, 55–66. Burlington, VT: Ashgate.

Hall, Colin M., Liz Sharples, Ritchard Mitchell, Niki Macionis, and Brock Cambourne. 2003. *Food Tourism Around the World: Development, Management and Markets*. Oxford: Butterworth Heinemann.

Henson, Spencer, and John Humphreys. 2010. "Understanding the Complexities of Private Standards in Global Agri-Food Chains as They Impact Developing Countries." *Journal of Development Studies* 46, no. 9: 1628–46.

Howkins, John 2002. *The Creative Economy: How People Make Money from Ideas*. London: Penguin.

Ilbery, Brian, Carol Morris, Henry Buller, Damian Maye, and Moya Kneafsey. 2005. "Product, Process and Place: An Examination of Food Marketing and Labelling Schemes in Europe and North America." *European Urban and Regional Studies* 12, no. 2: 116–32.

Liu, Shuwen, and Lewis T. O. Cheung. 2016. "Sense of Place and Tourism Business Development." *Tourism Geographies* 18, no. 2: 174–93.

Lockie, Stewart. 2001. "Food, Place and Identity: Consuming Australia's Beef Capital." *Journal of Sociology* 37, no. 3: 239–55.

Long, Lucy, ed. 2004. *Culinary Tourism*. Lexington: University of Kentucky Press.

Marques, Lenia, and Carla Borba. 2017. "Co-Creating the City: Digital Technology and Creative Tourism." *Tourism Management Perspectives* 24: 86–93.

Pantzar, Mika, and Elizabeth Shove. 2005. *Manufacturing Leisure: Innovations in Happiness, Wellbeing and Fun*. Helsinki: National Consumer Research Centre.

Pliner, Patricia, and Karen Hobden. 1992. "Development of a Scale to Measure the Trait of Food Neo-Phobia in Humans." *Appetite* 19, no. 2: 105–20.

Richards, Greg. 2009. "Tourism Development Trajectories—from Culture to Creative?" *The Asia-Pacific Creativity Forum on Culture and Tourism* 6: 9–15.

———. 2011. "Creativity and Tourism: The State of the Art." *Annals of Tourism Research* 38, no. 4: 1225–53.

Richards, Greg, and Julie Wilson. 2006. "Developing Creativity in Tourist Experiences: A Solution to the Serial Reproduction of Culture?" *Tourism Management* 27, no. 6: 1408–13.

Rudy, Jill T. 2004. "Of Course, in Guatemala, Bananas Are Better." In *Culinary Tourism*, edited by Lucy Long, 131–56. Lexington: University of Kentucky Press.

Schnell, Steven. 2011. "The Local Traveler: Farming, Food, and Place in State and Provincial Tourism Guides, 1993–2008." *Journal of Cultural Geography* 28, no. 2: 281–309.

Seyfang, Gill. 2006. "Ecological Citizenship and Sustainable Consumption: Examining Local Organic Food Networks." *Journal of Rural Studies* 22, no. 4: 383–95. https://doi.org/10.1016/j.jrurstud.2006.01.003.

Shenoy, Sajna. 2005. "Food Tourism and the Culinary Tourist." PhD diss., Clemson University.

Slocum, Susan L. 2015. "Local Food: Greening the Tourism Value Chain." In *Tourism in the Green Economy*, edited by Maharaj Vijay Reddy and Keith Wilke, 242–54. London: Routledge.

Slocum, Susan L., and Kynda Curtis. 2017a. "Farm Diversification through Farm Shop Entrepreneurship in the UK." *Journal of Food Distribution Research* 48, no. 2: 35–51.

———. 2017b. *Food and Agricultural Tourism: Theory and Best Practice.* London: Routledge.

United Nations World Tourism Organization (UNWTO). 2012. *Global Report on Food Tourism.* Madrid: UNWTO.

Urry, John. 2001. *The Tourist Gaze.* London: Sage.

Weatherell, Charlotte, Angela Tregear, and Johanne Allinson. 2003. "In Search of the Concerned Consumer: UK Public Perceptions of Food, Farming and Buying Local." *Journal of Rural Studies* 19, no. 2: 233–44.

Literary Atlas: A Digital Resource for Creative Tourism in Wales

Kieron Smith, Jon Anderson, Jeffrey Morgan

Recent developments in the tourism industry, as well as in the academic study of tourism, have resulted in new engagements with the "creative turn." Richards and Wilson (2006) trace this trend to the wider socio-economic context of late capitalism, and the postmodernist and "postmaterialist" attitudes that have arisen from this, wherein broad patterns of consumption have shifted from a focus on the consumption of objects toward a focus on the consumption of activities and experiences. Whereas social "distinction" (Bourdieu 1984) was once signalled by the things one *owns*, in recent years, distinction is increasingly signalled by the things one *does*. In the context of tourist activity, this has translated into a shift from the "spectatorial" goal of consuming as many locations as possible in one lifetime, that which Urry and Larsen describe as the "collecting of different signs that have been very briefly seen in passing at a glance" (2011, 20), toward more interactive forms of touristic "experience" (Pine and Gilmore 1999). No longer is the hackneyed photograph of the Taj Mahal or the Eiffel Tower taken at a distance enough to satisfy tourists possessing a more critical understanding of these "serially reproduced" signifiers of place (Richards and Wilson 2006). The "experience hunger of postmodern consumers" (Richards 2011, 1229) is such that tourists are increasingly seeking not simply to tick destinations off their bucket lists, but to engage in experiences through which they can inform, enrich, and express their senses of themselves.

The benefits of this state of affairs for tourist destinations in smaller communities are, potentially, numerous. For one, the shift in appetite from the consumption of tangible destinations to "intangible cultural resources" means that those locations that do not possess a "rich built heritage"—Taj Mahals and Eiffel Towers—can potentially attract tourists on the basis of their intangible cultural heritage (Richards 2011, 1230). As Duxbury notes in chapter 1 of this volume, creative tourism can attract visitors on a more manageable, smaller scale, and outside the "usual high-tourism season." Moreover, smaller communities, by virtue of their smallness, tend to be "distinctive" in ways that larger destinations are not. In an era of intense competition for tourist footfall, smaller communities can offer tourists unique and unusual experiences, giving them a "symbolic edge" (Richards 2011, 1230) over their more recognizable competitors. Furthermore, a touristic interest in those intangible, unquantifiable features that make a place "distinctive"—landscapes, languages, cultures, literatures—means that these things might be celebrated and safeguarded for future generations in ways they may not ordinarily be. Creative tourism provides creative people within smaller communities additional revenue for their skills, giving them more opportunities to *create*, and thereby develop and sustain their own practices, resources, and collaborative networks (see Duxbury in this volume). Combined with the provision of "creative" participative experiences, in which tourists are invited to become "co-performers and co-creators" in experiences that enable "self-realization and self-expression" (Richards 2011, 1237), it may therefore be possible to preserve and develop cultural distinctiveness, providing resources for future inhabitants while at the same time attracting tourist trade.

However, there are challenges. Tourism is a powerful part of the "cultural and symbolic economy" (Richards 2011, 1237) the world over, one of the key mechanisms through which culture is produced and reproduced. Given the onus the OECD places on destinations to "transform the basic inherited factors into created assets," to "make better use of their inherited and created assets to make themselves attractive to tourists" (2009, 29–30), there is the danger that cultural inheritances are reshaped and commodified in the image of the priorities of mass tourism. In this context, smaller communities, again by virtue of their smallness, are arguably more vulnerable to such economic and cultural pressures than larger destinations.

Wales is a case in point. With a population of little over 3 million, Wales is by most measures a "small nation." Though situated within the United Kingdom, since the formation of a devolved administration in 1999, it has possessed a measure of political independence. A stagnating post-industrial economy has resulted in a concerted effort by the Welsh Government (WG) to push toward the promotion of Wales as a tourist destination. One corollary of this has been an emphasis in policy literature on "distinctiveness," and on offering tourists a "strong Welsh experience" (Visit Wales 2011, 1). Meanwhile, the Well-being of Future Generations (Wales) Act 2015, followed by the creation of a Welsh "future generations commissioner" in 2016, placed the onus on the WG to secure a sustainable ecological, social, and cultural environment for future generations. The question for tourism development in the Welsh context, then, is whether tourism can be harnessed—as it has been so successfully in, for example, Iceland (see Aquino and Burns in this volume) and Portugal (Duxbury in this volume)—as a means of improving the economy whilst simultaneously preserving and developing the nation's distinctiveness in an ecologically, socially, and culturally sustainable way.

This is the context in which *Literary Atlas* has been developed. *Literary Atlas: Plotting English Language Novels in Wales* is an innovative, interdisciplinary exercise in the field of literary geography.[1] The project has created an interactive website showcasing digital deep maps of a selection of novels written in the English language and set in Wales. Alongside this, it has designed an evaluative framework, organizing reading groups and field trips to locations around Wales, encouraging participants—including tourists—to engage with Welsh space and place through the lens of literature, while at the same time evaluating these engagements as potential uses of the *Literary Atlas* website. This chapter will outline the main features of the project and examine some of its findings. It will question whether digital deep mapping of this kind can offer a resource through which residents and visitors can engage with Welsh space and place in sustainable and mutually beneficial, co-creative ways.

Tourism has been an important feature of the Welsh economy for some time. Wales's economy was once dominated by the extraction of natural resources—primarily, and most famously, from the late nineteenth century onwards, "King Coal" (Morgan [1981] 1998, 322). By the early twentieth century, Wales was a global centre of heavy industry, but with the decline in the profitability of coal in the 1930s, and the decline in the profitability of those heavy industries that replaced it in the 1980s, Wales's economy has in recent decades been restructured around the service and creative industries, with a major focus on tourism. By the mid-1990s, tourism accounted for nearly one in ten Welsh jobs and "contributed as much to the gross domestic product of Wales as it did in Greece or Spain" (Johnes 2012, 403). Moreover, following the global economic crash of 2008, further hopes were pinned on tourism as a means to attract international investment. As one WG strategy document notes:

> Tourism has a key role to play in reinforcing a distinctive and compelling national identity for Wales in the UK and Internationally as a place to visit, invest in and as a place to do business. It can help to reflect our true culture and character. . . . Perceptions of Wales are influenced by a complex mix of factors. Not all of these can be overcome easily but consistent and reinforcing branding can help to overcome weak associations with Wales and develop positive attitudes. (Welsh Government 2013, 7)

Indeed, the tourist industry is an invaluable component of the Welsh economy. Not only is it a major employer that continues to support around 10 per cent of the Welsh workforce, the sector also supports jobs across the spread of the country (Welsh Government 2013, 7), while other sectors—public and private services, the media, finance—are concentrated in South Wales, particularly in the capital, Cardiff. Under the devolution settlement in place at the time of writing, the WG possesses no direct revenue-raising powers; however, it does have the remit to devise its own distinct Welsh tourism strategy as a means to stimulate domestic and international tourism and foreign direct investment. Since 2005 this function has been operated by a department within the WG, Visit Wales.

In the context of what it admitted were the "exceptional financial challenges" of the post-crash era (Welsh Government 2010, 3), compounded by increased competition in both domestic and international tourism markets (Welsh Government 2013, 5), the WG has in recent years pursued a tourism strategy that focuses on developing Wales's unique "brand" for domestic and international markets and shaping Wales as a "distinctive" destination. To this end, it has taken up the widespread enthusiasm for "mega-events" in Western economies (Richards and Palmer 2010; Rojek 2013). In 2010, the WG established a Major Events Unit to attract and create mega-events in Wales. This new unit was set a remit to create "a positive external reputation and brand image for Wales" and to "deliver a series of economic, social, cultural, and legacy benefits" (Welsh Government 2010, 3). These would primarily consist of major sporting events, such as the 2010 Ryder Cup (Harris 2015), but would also include musical and cultural festivals (Welsh Government 2010). In the area of culture, the Major Events Unit opted to use the recent centenaries of two internationally recognized writers of Welsh birth, Dylan Thomas (2014) and Roald Dahl (2015), as the basis for year-long cultural festivals celebrating the legacy of these writers and, naturally, capitalizing on their value as well-known figures. Following the perceived success of these events, Visit Wales has since pursued a "thematic years approach" to tourism in Wales. The WG declared 2016 as the "Year of Adventure," 2017 the "Year of Legends," and 2018 the "Year of the Sea" (Welsh Government 2016), with the emphasis being the "long-term ambition to grow a stronger and more defined brand for tourism in Wales" (Welsh Government 2016).

This broad-brush "thematic" approach to tourism draws upon research that suggests that the creation of "distinctive" national branding plays an important part in attracting foreign direct investment and, of course, tourists. Visitor research conducted by Visit Wales showed that "having a distinct, authentic Welsh experience is an important factor in influencing trip satisfaction" (2012, 1). Indeed, the WG tourism policy and strategy documents since 2008 have been peppered with the terms "authentic" and "distinctive," and the two are often used together, interchangeably, as the following excerpts show: "travellers are looking for distinctive, authentic experiences" (1); "having a distinct, authentic Welsh experience" (1); "the presentation of authentic, compelling and distinctive stories" (6); "Authentic experiences: We need to provide a distinctive sense of place experience" (Welsh Government 2013, 27). However, "distinctiveness" and "authenticity" are, of course, not synonymous terms. It is possible to be "distinctive" without being remotely "authentic," and vice versa. Inevitably, these conceptual slippages spill out into the practice of cultural tourism itself; indeed, it could be argued that in recent years, cultural tourism in Wales has begun to emphasize "distinctiveness" at the expense of "authenticity." Of course, after post-structuralism, "authenticity" is itself a fraught, problematic concept. But in terms of national culture, we can perhaps settle on a definition that rests in some sense on the notion of culture that has been produced by individuals and communities of a given nation, or culture that responds in some way to "places" and the individuals and communities that inhabit them. Yet in recent years this seems to have been sidelined in favour of "distinctiveness." For instance, in 2016 Wales saw a major centenary celebration of the birth of writer Roald Dahl, which was that year the main focus for Literature Wales, the organization tasked with overseeing Wales's literary heritage. That year witnessed major exhibitions, adaptations, and a major two-day carnival event in Cardiff—City of the Unexpected—funded and promoted by the WG through Literature Wales. Roald Dahl was of course an indisputably brilliant writer, and in particular the City of the Unexpected event was, in tourism terms, a "distinctive" success, bringing thousands of people to Cardiff (Price 2016). But the question of whether this was a celebration of an "authentic" Wales is open to debate. Dahl was born in Cardiff to Norwegian parents, but at age eight left to be educated in the English public school system; thereafter, he never returned for longer than brief holidays. In a context in which tourism is an indispensable facet of a fragile Welsh economy, an approach to tourism that is able to showcase both the "distinctive" and "authentic" aspects of Welsh life and culture seems preferable.

It should be said that Welsh tourism is not the primary analytical focus of the *Literary Atlas* project. However, the conceptual framework within which the project operates overlaps in fruitful ways with recent developments in tourism studies. The term "literary geography" was famously first used by William Sharp in his 1904 book of the same name, which attempted to offer an account of what he called "the distinctive features of the actual or delineated country of certain famous writers, and on certain regions which have many literary associations" such as "Dickens-land" (South East England), "Scott-land" (Scotland), or "The Literary Geography of the Lakes" (Cumbria) (1904, vii). This was an approach that viewed places and landscapes as static backdrops to literary works. However, the theoretical conceptualization of the field has since shifted considerably. Literary geography is informed by the intersection of the spatial turn in the humanities and the cultural turn in the social sciences that has taken place over the past few decades. In both the humanities and social sciences it is now commonly understood that "places" are not stable, fixed backdrops to human experience, straightforwardly available to representation, but plural, processual, and culturally encoded: experienced and created through human activity. In Edward Soja's terms, the human experience of place is best understood as hyphenated, dynamic, "real-and-imagined" (1996). Literary geography is therefore an interdiscipline concerned with the place of the literary imagination within these complex, dynamic processes of place creation—in Franco Moretti's words, the sense in which works of literature are a part of the "ongoing composition" through which "real-and-imagined" places are made and lived (1998, 35).

Literary Atlas is primarily an exploration of these varied, complex processes of place creation and composition. Its aim is to establish a new way of mapping English-language novels set in Wales in order to gain fresh insights into the role of "place" in the relationship between who we are as individuals and groups, and the geographical locations that shape these identities. To this end, the project has set out to create an online, open-access digital atlas of literary cartographies of twelve English-language novels set in Wales. In a key sense, the project aims to use mapping as a way to embark on a Deleuzian "deterritorialization" of the dominant "cognitive maps" (Jameson 1987) of the literature of the British Isles, which often overlooks writing from the smaller community of Wales. This is clearly an important task for a small nation in an era in which cultural and creative tourism are on the rise. In this sense, digital mapping is a powerful means of enabling a destination such as Wales to harness, in Richards's terms, "the aspects of creativity that are linked to place" (2011, 1238) in a way that is sensitive to the particular cultural and economic needs of that place.

Selecting the twelve novels was no easy task. There is a long and rich tradition of English-language writing in Wales, and it took some time to whittle down a lengthy longlist of possibilities to just twelve. They were ultimately selected according to the following criteria:

- The narrative must be based in Wales, with a specific Welsh place or region as central to the narrative.

- The book must be written in English. (There is a rich and distinct anglophone literary tradition worth studying in its own right. This criterion is not meant to downplay the wealth of Welsh-language literature in print. Were this to be a comprehensive atlas of literary Wales, literature in both languages would need to be included).

- The books must be works of fiction that take the form of the novel. This was a pragmatic scoping decision in order to provide a clear focus for the *Literary Atlas*. There is a plan to expand the site to plays, poems, short stories or non-fiction in the future.

- The novels should offer a cross-section of genre, and represent different eras in Wales's anglophone literary history.

- The novels should offer as broad a geographical coverage as possible.

- The novels should represent a broad diversity of authors active in Wales (in relation to gender, ethnicity, age, etc.). The book does not have to be by a Welsh-born author.

- The final list should include both "classic" Welsh fiction and contemporary writers in order to draw attention to new materials in addition to established works.

From these criteria, the *Literary Atlas* team consulted with a range of academic and literary experts in Wales, as well as asking the public to offer their own choices for selection. The final shortlist maps literature from a broad range of Wales's social and physical geographies:

1. Raymond Williams's *Border Country* (1960). A historical novel charting the lives of the Prices, a family living in the village of Pandy, near Abergavenny, close to the Welsh-English border, from the 1920s to the 1950s.

2. Malcolm Pryce's *Aberystwyth Mon Amour* (2001). A fantastical parody of the hard-boiled crime novel, set in a fictionalized Aberystwyth, a coastal town on the scenic Ceredigion coastline.

3. Alys Conran's *Pigeon* (2016). A story of two young teenagers growing up in and around Bethesda, a post-industrial slate-mining town in Snowdonia.

4. Tristan Hughes's *Revenant* (2008). Three young teenagers return to the site of a collective trauma near their hometown; the place is unnamed but based on the pretty coastal town of Beaumaris.

5. Christopher Meredith's *Shifts* (1988). A novel recounting the social and emotional impact of the slow closure of the colossal Ebbw Vale steelworks on four main characters.

6. Menna Gallie's *Strike for a Kingdom* (1959). A "whodunit" set in a small South Wales mining community of Ystradgynlais during the 1926 general strike.

7. Trezza Azzopardi's *The Hiding Place* (2001). The story of the Gaucis, a Maltese family living in the multicultural Butetown area of Cardiff—known colloquially as Tiger Bay—in the 1960s.

8. Alan Garner's *The Owl Service* (1967). A young-adult fantasy novel that reimagines the fourth branch of the *Mabinogion*, set in the beautiful yet remote village of Llanymawddwy, in North Wales.

9. Amy Dillwyn's *The Rebecca Rioter* (1881). A romping historical novel recounting a young man from a village just outside Swansea caught up in the Rebecca Riots of the 1840s.

10. Niall Griffiths's *Sheepshagger* (2001). The tragic tale of Ianto, a troubled young native of an unspecified location somewhere in the Cambrian mountains of West Wales, caught up in the drug counterculture of the 1990s.

11. Fflur Dafydd's *Twenty Thousand Saints* (2008). A novel exploring notions of Welsh cultural and political identity through the experiences of a group of characters staying on Bardsey Island one summer.

12. Lloyd Jones's *Mr Vogel* (2004). A fantastical travelogue of a perambulatory journey around the entire coast of Wales.

Beyond the de/reterritorializing potential of *Literary Atlas*, a key feature of the project is its use of "deep mapping" as a means to unpack the multi-faceted, processual, palimpsestic nature of places. Deep mapping is a malleable concept that has been interpreted in innumerable ways by cartographers, artists, and academics. Its origins lie in the work of Guy Debord and the Situationist movement, but its more recent articulations can usefully be traced to the post-structuralist problematization of the notion of objective, "Cartesian" mapping as an adequate means of representing spaces and places. Deep maps, by their nature, are designed to disrupt and destabilize, in Doreen Massey's terms, the "elitist, exclusivist enclosures within which so much of the production of what is defined as legitimate knowledge still goes on" (2005, 75). One potential function of the deep map is therefore to destabilize normative, "spectatorial" knowledges of place—knowledges that, in the search for "distinctiveness," can omit or occlude the complex forces that shape places and those who inhabit them—and invite more nuanced, "creative" experiences and inhabitations of place. Deep mapping the ways in which works of literature inhabit places is one means of exploring and unpacking these multiple layers, and opening up new ways of understanding and experiencing place.

Literary Atlas has pursued an approach to digital deep mapping formulated by Damian Walford Davies, who posited the idea of a "Digital Literary Atlas of Wales" in his 2012 book *Cartographies of Culture*. This is an approach to literary mapping that, in Walford Davies's terms

> Resist[s] simply "imposing" cartographic templates on literary works, and merely extracting various kind of mappable data from them. Rather, by offering the user/reader an interactive platform on which to explore a variety of cartographic

and geographical "contexts" in relation to which a given author and literary work can meaningfully be located, the . . . [atlas prompts] a critical and affective inhabitation of the cultural dimensionality of a literary work. (2012, 206)

In other words, the team started with "extractions" of "mappable data"; in our case, we geoparsed every reference to a "real" or approximated place contained in each of the twelve novels, and named these "plotpoints." However, this was done with the aim of exploring the wider "cultural dimensionality" of those works. This "cultural dimensionality" refers to the broad, potentially endless range of geographical, historical, and cultural references, intertexts, and textures that constitute "real-and-imagined" place. Our deep maps therefore augment these plotpoints with a range of supplementary materials, such as selected extracts from the chosen novels; excerpts from interviews undertaken with authors and experts; audio extracts from the novels read by the author and the *Literary Atlas* team; street-view images of the novels' locations; historical and contemporary maps of the novels' locations; photos and films connected to the sites of the novels; and commentary from the *Literary Atlas* team.

These deep maps can be navigated independently by the *Literary Atlas* user, but they are also curated into routes, which we have termed "plotlines." These plotlines are suggested routes through the literary and geographical landscape of the novel, its "cultural dimensionality," narrating insights into the "plot" of both text and place. Anderson has described plotlines as both "geographical and literary routes through the world. They are cartographic navigations that connect the page to the place and entangle them into an ongoing composition" (2016, 161). In other words, plotlines are physical routes that enable users to explore the entanglements of the "real" and "imagined." The *Literary Atlas* user can choose a plotline for a particular novel and scroll through features of its deep map to gain a new understanding of the relations between literature and geography. Sometimes a plotline will follow a route a character has followed; for instance, one plotline for *The Rebecca Rioter* traces the route taken by the protagonist, Evan Williams, on his escape from the authorities, having taken part in the Rebecca Riots. Sometimes a plotline will trace a route through a key thematic feature of a novel; one plotline for the novel

Sheepshagger pursues the ways the novel explores the notion of "wildness" in contrast to its depictions of urban space. Other times, a plotline will explore an important aspect of an author's writing process; one plotline traces the biography of author Alan Garner in the years leading up to his writing of *The Owl Service*. All of these plotlines attempt to shed new light on the complex ways in which physical and social place play their part in the literary imagination. It is possible to follow these plotlines on a computer, tablet, or smartphone in any location—enabling the user to engage with these routes virtually, as well as, crucially, to walk them in practice. We will return to this point shortly.

Beyond the twelve showcased deep maps, the website provides a range of other scholarly and creative resources. Some of these may be useful for students and scholars of the literary geography of Wales, such as various visualizations of the aggregated data, and bibliographies of secondary literature. Others we anticipate will be fruitful resources for the creative tourist. One of these is a comprehensive "library" map of all anglophone novels set in or connected to Wales. Users can view an interactive, searchable map of Wales and discover novels organized by location. The first iteration of this map was created using the library catalogues of Swansea University, Cardiff University, and the National Library of Wales; all English-language novels collected in the "Wales" collections of these libraries were scanned, and a single geolocation attributed to each novel, based on an approximate general region in which the novel is set, or a central location in the novel's imagined geography. Currently the map contains over three hundred books, though this number will increase with the development of an interactive function, whereby users can add to the map themselves.

However, the broader aim of *Literary Atlas* is not only to provide users with an informational resource, but also with a practical resource that encourages and enables new creative interactions with Wales's varied physical and cultural geographies. While the plotlines and library maps on the website enable users to embark on experiences of place filtered through the lens of literature, they equally invite users to engage with place in a way that moves beyond passive, spectatorial experience, and toward something more active, dialogic, embodied, and creative. One of the key features of the site in this regard is an interactive "microfiction" map, whereby users can write their own five-hundred-word microfictions inspired by Welsh locations and post these online for other users to read. *Literary Atlas* defines microfictions as any brief creative reflection on a place; these may be short narratives, fictional vignettes, travelogues, poetry, creative non-fiction, or any other form the user-writer wants to employ. The interactive interface provided on the site enables users to easily select any geographical location in Wales or the world on which to pin their microfiction, as well as upload a selection of photographs to accompany their work. Set alongside the maps of published fiction set in Wales, these microfictions will allow visitors to the website and to Wales to contribute creatively to the "ongoing composition" of Welsh places.

Furthermore, a key aim of the *Literary Atlas* project is to encourage users to visit literary places for themselves. As a means of exploring and evaluating the possibilities of such uses of the *Literary Atlas* website, the project organized a series of literary field trips to locations from the twelve novels listed above. Informed by recent scholarship on the efficacy of walking as a research tool (Anderson 2004; Ingold and Vergunst 2008; Bates and Rhys-Taylor 2017), these trips operated under the understanding that places are not "passive stages in which actions occur," but rather "the medium that impinge on, structure and facilitate these processes" (Anderson 2004, 255), and that interviewing participants in situ might be a powerful means to uncover the complex, elusive relations between people and place. As Anderson notes, "talking whilst walking" offers the possibility to "tap into the non-mechanistic framework of the mind and its interconnections with place to recall episodes and meanings buried in the archaeology of knowledge" (2004, 260). With these reflections in mind, the project invited participants to take part in field trips to the "real," physical locations in which the twelve books are set. During these excursions we took journeys through the "literary spaces" (Saunders and Anderson 2014). Beyond reminding participants of the novels' plots, we did not provide further historical or contextual information about the locations; rather, we stopped to read passages from the novels in situ in order for participants to respond in their own way to the literary representation of the spaces. We recorded participants' responses while on location, as well as in discussions after the event.

Visiting a place or walking a route described in or determined by a literary narrative may be understood as a form of "experience" tourism (Pine and Gilmore 1999). Indeed, if, as Rebecca Solnit has argued, "narrative writing is closely bound up with walking" because, "just as with following footsteps, it allows one to read the words of someone—the author—who has gone before" ([2001] 2014, 72), then it follows that following the footsteps of an author or character can offer the opportunity to walk in the footsteps of their "experience." This notion of "experiencing" a place through the eyes of a character was confirmed in one trip to a key location from Christopher Meredith's novel *Shifts*. *Shifts* explores the social and emotional effects of the gradual closure of the Ebbw Vale Steelworks—a major industrial employer in the South Wales Valleys—at the end of the 1970s, and the different ways in which four characters cope with the changes this protracted closure makes to their lives. One character, Keith, copes by attempting to connect with the industrial history of his hometown, a fictionalized version of the real town of Tredegar. Throughout the course of the novel, Keith visits key locations from the history of the town in his effort to better understand who he is and where he is from. On the *Literary Atlas* field trip to Tredegar, we visited one of these sites, the ruins of the eighteenth-century Sirhowy Ironworks, and read a passage from the novel with the participants. This is one section of that passage:

> [Keith] tramped down to the ruin and stared at it, conscious of his looking, like a viewer at an art gallery wondering what to think. It was like repeating a word till it becomes meaningless. He ran a hand over the stone and a crumbling mortar, trying to concentrate. There was still an overhanging plug of iron slag where some stonework had fallen away. The remains of the last tapping. He moved around the ruin running his hand under the overhang and spoke again to himself. (Meredith 1988, 18)

Keith has, in effect, embarked on a field trip of his own, and this enabled *Literary Atlas* to follow his footsteps and experience his self-conscious visit to a local site of historical interest in a quite literal way. Interestingly, the participants' response to the location corresponded with Keith's sense of bewilderment. In the passage above, though Keith knows what he is looking at, he does not quite know "what to think." Similarly, one participant expressed a similar sense of incomprehension:

> When you go to museums they frame it for you, don't they? They tell you why it's important and why you're here. . . . People go because they want to see this important piece of whatever because it's celebrated that way, but . . . I don't know what I'm supposed to think about it. (*Literary Atlas* field trip, 19 October 2017)

This sense of incomprehension did not appear to impair this participant's enjoyment of the field trip; rather, it seemed to enable a more nuanced understanding of the point of view of the character:

> I can understand why. . . . I think he's hit it in a nutshell saying that. What do I think when I look at this? And I suppose if you go into an art gallery and it's art that, you know, like minimalism, which many people don't take to, you know, you think what the hell am I supposed to think of this, you know? . . . And I can understand, I think he's hit it on the head there, I hadn't noticed that phrase, but I think he's hit it on the head there. (*Literary Atlas* field trip, 19 October 2017)

In effect, the act of following in the footsteps of a character from the book enabled the participant to acquire not only an experience of the character's point of view, but also, it appears, a stronger understanding of both the book and the location, as well as perhaps the way in which the book responds to and contributes to the "ongoing composition" of the location.

This particular section of *Shifts* was an unusually appropriate one to pursue on a field trip, given that Keith is on a field trip of his own. More often, the *Literary Atlas* field trips would explore locations depicted in a less self-conscious way. However, where possible, the trips did visit locations that are not only described but reflected on by a character or narrator. Another example was on a trip taken to the town of Beaumaris on the island of Anglesey (Ynys Môn), off the coast of North Wales, one of the main settings of Tristan Hughes's novel *Revenant*. One of the locations visited on this trip was the iconic "West End," a terrace of Georgian-style houses well-known due to its striking and somewhat incongruous architecture and its prominent location on the main approach into the town. The houses are colourfully painted and feature on innumerable tourist images of the area: postcards, guidebooks, souvenirs. We visited this location with a group of participants, and read the following passage narrated by the jaded character of Ricky, who has reluctantly returned to his hometown after a spell working in England:

> I park the car on the West End, in front of the Georgian terrace, and I'm sure they've given the houses a lick of paint or five since I last saw them 'cause they're positively beaming today, glowing like, even though it's April and the sun isn't so strong yet. They've done them in rose-pink and lime-green and primrose-yellow, and other colours too, as if they wanted to turn all this stone and concrete into coral beds, to flip the world upside down and have the underwater colours on top, in the air. I don't get it with these seaside places—the way they chuck their slap on when hardly anyone below a hundred and ten wants to go to them any more, like some old bird plastering herself when she knows the best she's going to do is get groped by some pensioner at the bar. I mean, who do they think they're going to fool? Please, come here, we're such jolly folk we paint our houses happy colours. Please stay. Please fork out sixty quid for a night in a place that looks like a frigging rainbow on the outside but is still damp and grim and grey on the inside. A word to the wise: the brighter and more colourful these places look, the more fucking *desperate* they are. (Hughes 2008, 27)

Here the novel offers an interpretation of the location that runs against the grain of the intended tourist gaze. Ricky, a native, is decidedly cynical about the touristification of his hometown, having seen its less palatable features growing up as a teenager there. In keeping with Anderson's contention that on-location interviews allow participants to "recall . . .

meanings buried in the archaeology of knowledge" (2004, 260), the passage was met with strikingly different responses by participants possessing different ideas about the location. Two of the participants who had never visited Beaumaris before appeared to agree with Ricky's view, and brought their experiences of their respective hometowns in the Netherlands and Germany to bear on this. The German participant suggested that while they could appreciate the attractiveness of the location, they could agree there were other ways to see it: "To me, it looks fine. It looks nice, but I do get it. If I would have been from here, I think I would have had pretty much the same opinion, very similar opinion [to Ricky]. If I compare it to my town, I feel very much like this about my hometown" (*Literary Atlas* field trip, 23 September 2017). Yet two participants who, though they settled in the area as adults, live on Anglesey and regularly visit Beaumaris, responded somewhat differently. Both disagreed strongly with Ricky's jaundiced view, with one stating after the passage had been read that "It says more about him than it does the terrace, I think" (*Literary Atlas* field trip, 23 September 2017). Another stated they thought the painted houses were less about attracting tourism and more a reflection of local pride: "Because the people have got pride and they want their village nice" (*Literary Atlas* field trip, 23 September 2017). However, later in the day, during the post-trip discussion, one of these participants appeared to have changed their mind, and having visited the town through the eyes of the novel, decided they could see Ricky's point of view: "It just made me think that if I lived in Beaumaris and actually experienced it day to day and didn't live in one of the picturesque painted houses and just lived in around the back roads, whether I would still have the view, the glowing view that I have of it because of the way I visit it" (*Literary Atlas* field trip, 23 September 2017).

Anderson has argued that reading fictions in the context of the "real" places in which they are set can "help us tell a new truth about the relations between people and place." Examining the "ongoing compositions" between text and place, he argues, "we can locate our own lives, finding new meanings which resonate with our own experiences and broaden our horizons" (2014, 314). The experiences of field trip participants explored here suggest that visiting locations—whether as a tourist or a resident—through the lens of a work of literature can powerfully inform one's perspective on those places, as well as one's understanding of the work itself.

It is hoped that the range of digital resources available on the *Literary Atlas* website—the deep maps of selected novels, the distant "library" map, the "microfictions" feature, and much else besides—will prove to be a fruitful resource for users seeking not only "spectatorial" engagements with "serially reproduced" (Richards and Wilson 2006) Welsh place and culture, but also creative experiences that enable, in Richards's words, "self-realization and self-expression" (2011, 1237). While there are innumerable spectatorial opportunities available in Wales—the sublime heights of Snowdonia in North Wales, the sweeping beaches of the Gower Peninsula in the South, and plenty more in between—it is hoped that *Literary Atlas* will prove to be a resource for deeper and more creative authentic experiences of the rich multiplicity of Welsh places. As Scherf notes in the introduction to this collection, cultural mapping can provide a way of enabling places to showcase their "unique fingerprint on the planet." Creative literary tourism that utilizes such maps can provide the basis for throwing into relief those "intangible" cultural resources off the beaten track, while at the same time being sensitive to the diverse histories and stories of those places. Seeing these locations through the lens of literature can open up touristic experiences that contribute to the "ongoing composition" of places, offering new perspectives and co-creating sustainable new experiences and senses of place for today and the future.

NOTES

1   The project's website is available at www.literaryatlas.wales.

References

Anderson, Jon. 2004. "Talking Whilst Walking: A Geographical Archaeology of Knowledge." *Area* 36, no. 3: 254–61.

———. 2010. *Understanding Cultural Geography.* London: Routledge.

———. 2014. *Page and Place: Ongoing Compositions of Plot.* Amsterdam: Rodopi.

———. 2016. "Between 'Distant' and 'Deep' Digital Mapping: Walking the Plotlines of Cardiff's Literary Geographies." In *Literary Mapping in the Digital Age*, edited by David Cooper, Christopher Donaldson, and Patricia Murrieta-Flores, 161–79. London: Routledge.

Bates, Charlotte, and Alex Rhys-Taylor, eds. 2017. *Walking through Social Research.* Abingdon, UK: Routledge.

Bourdieu, Pierre. 1984. *Distinction: A Social Critique of the Judgement of Taste.* Translated by Richard Nice. London: Routledge.

de Certeau, Michel. 1984. *The Practice of Everyday Life.* Berkeley: University of California Press.

Harris, John. 2015. "Keeping Up with the Joneses: Hosting Mega-Events as a Regenerative Strategy in Nation Imaging, Imagining and Branding." *Local Economy: The Journal of the Local Economy Policy Unit* 30, no. 8: 961–74.

Hughes, Tristan. 2008. *Revenant.* London: Picador.

Ingold, Tim. 2011. *Being Alive: Essays on Movement, Knowledge and Description.* London: Routledge.

Ingold, Tim, and Jo Lee Vergunst, eds. 2008. *Ways of Walking: Ethnography and Practice on Foot.* London: Routledge.

Jameson, Frederic. 1987. "Cognitive Mapping." In *Marxism and the Interpretation of Culture,* edited by Cary Nelson and Lawrence Grossberg, 347–60. Chicago: University of Illinois Press.

Johnes, Martin. 2012. *Wales Since 1939.* Manchester: Manchester University Press.

Jones, Calvin. 2005. "Major Events, Networks and Regional Development." *Regional Studies* 39, no. 2: 185–95.

Massey, Doreen. 2005. *For Space.* London: SAGE.

Meredith, Christopher. 1998. *Shifts.* Bridgend, UK: Seren.

Moretti, Franco. 1998. *An Atlas of the European Novel, 1800–1900.* London: Verso.

Morgan, Kenneth O. (1981) 1998. *Rebirth of a Nation: A History of Modern Wales.* Oxford: Oxford University Press.

OECD. 2009. *The Impact of Culture on Tourism.* Paris: OECD.

Pine, Joseph P, and James H. Gilmore. 1999. *The Experience Economy.* Boston: Harvard Business Review Press.

Price, Karen. 2016. "Thousands Turn out to Watch Roald Dahl's City of the Unexpected Come to Life." *WalesOnline,* 17 September. https://www.walesonline.co.uk/whats-on/family-kids-news/thousands-roald-dahl-city-unexpected-11899696.

Richards, Greg. 2011. "Creativity and Tourism: The State of the Art." *Annals of Tourism Research* 38, no. 4: 1225–53.

Richards, Greg, and Robert Palmer. 2010. *Eventful Cities: Cultural Management and Urban Revitalisation.* London: Routledge.

Richards, Greg, and Julie Wilson. 2006. "Developing Creativity in Tourist Experiences: A Solution to the Serial Reproduction of Culture?" *Tourism Management* 27, no. 6: 1408–13.

Rojek, Chris. 1995. *Decentring Leisure: Rethinking Leisure Theory*. London: Sage.

———. 2013. *Event Power: How Global Events Manage and Manipulate*. London: Sage.

Saunders, Ahmed, and Jon Anderson. 2015. "Relational Literary Geographies: Co-Producing Page and Place." *Literary Geographies* 1, no. 2: 115–9.

Scherf, Kathleen. 2015. "Beyond the Brochure: An Unmapped Journey into Deep Mapping." In *Cultural Mapping as Cultural Inquiry*, edited by Nancy Duxbury, W. F. Garrett-Petts, and David MacLennan, 338–60. London: Routledge.

Sharp, William. 1904. *Literary Geography*. London: Pall Mall Publications.

Soja, Edward. 1996. *Thirdspace: Journeys to Los Angeles and Other Real-and-Imagined Places*. Oxford: Wiley-Blackwell.

Solnit, Rebecca. (2001) 2014. *Wanderlust: A History of Walking*. London: Granta.

Urry, John and Jonas Larsen. 2011. *The Tourist Gaze 3.0*. London: Sage.

Visit Wales. 2012. *Cultural Tourism Action Plan 2012–2015*. Cardiff: Visit Wales. https://businesswales.gov.wales/dmwales/sites/dmwales/files/documents/DM%20Wales%20-%20Cultural%20Tourism%20-%20Action%20Plan%20-%20Eng.pdf.

Walford Davies, Damian. 2012. *Cartographies of Culture: New Geographies of Welsh Writing in English*. Cardiff: University of Wales Press.

Welsh Government. 2010. *Event Wales: A Major Events Strategy for Wales, 2010–2020*. Cardiff: Welsh Government. https://gov.wales/sites/default/files/publications/2019-06/event-wales-a-major-events-strategy-for-wales-2010-to-2020.pdf.

———. 2013. *Partnership for Growth: Welsh Government Strategy for Tourism, 2013–2020*. Cardiff: Welsh Government.

———. 2016. "Thematic Years Approach." https://businesswales.gov.wales/tourism/thematic-years#:~:text=In%202015%2C%20we%20announced%20a,a%20series%20of%20annual%20themes.&text=The%20strategy%20is%20driven%20by,promoting%20Wales%20as%20a%20destination.&text=You%20may%20also%20like%20to,consumer%20site%20(external%20link).

Creative Tourism Opportunities through Film and Tourism Industry Collaboration

Christine Van Winkle, Eugene Thomlinson

Strong film and tourism industries that work in partnership have the potential to contribute positively to small communities and offer creative tourism opportunities that go beyond just film tourism. Independently, film and tourism activities can each raise the profile of a community, provide employment opportunities, and result in spinoff expenditures from legacies that remain within communities such as film sets and memorabilia (Wray and Croy 2015; Cardoso et al. 2017; Tkalec, Zilic, and Recher 2017). Mandic, Petric, and Pivcevic (2017) argue that creative industries such as film are not only interconnected with tourism, they form an essential element of a sustainable tourism industry. Film productions act as drivers of creative tourism by contributing to resource development, film tourism, and destination image enhancement, which can attract visitors to the destination long after the film has wrapped. Films can accomplish this by building narratives, presenting a unique identity, and contributing to the destination brand (Juskelyte 2016; Richards and Wilson 2006), thereby "offering regions and cities a symbolic edge in an increasingly crowded marketplace" (Richards 2011, 1230). And yet, unfortunately, these two industries often work in isolation, reducing the potential benefits for the local community.

While creative tourism is conceptualized in a variety of ways, definitions typically include a description of experiential tourism that includes sharing cultural and place-based experiences and engaging visitors in the creative aspects of the destination (Duxbury and Richards 2019). In the introduction to this volume, Scherf points out that "a collaborative paradigm in culture-led tourism" can contribute to creative tourism. This chapter, then, explores how collaboration between film and tourism industries can contribute to creative tourism opportunities at the destination.

Research examining film and tourism has focused on a range of topics, including location and destination marketing (Bolan and Williams 2008; Croy 2010; Hahm and Wang 2011; Hudson and Tung 2010; Volo and Irimais 2016; Ward and O'Regan 2009); destination image impacts (positive and negative) through film (Chen 2018; Hudson, Wang, and Gil 2011; Juskelyte 2016); the influence of film celebrities on destination image (Chen 2018); film-tourism product development (Carl, Kindon, and Smith 2007; Kim 2012); and the relationships between film and tourism industries (Tkalec, Zilic, and Recher 2017: Hudson 2011). There is less research examining collaboration between these industries and the need for planning to maximize benefits (Croy 2010). Leveraging film production and tourism initiatives to develop creative tourism that benefits communities requires planning and should involve both parties working in tandem. Richards (2011) states that "the spontaneous nature of much creative activity does not lend itself easily to planning" (1244), which can be problematic for strategic tourism development initiatives. The purpose of this chapter is to explore opportunities for collaboration between the film and tourism industries to benefit small communities through creative tourism planning and development.

Mandic, Petric, and Pivcevic (2017) discuss the importance and value that the cultural and creative industries provide to the entire economy, in addition to encouraging innovation. Creativity as an element of tourism has been examined in relation to both the tourism experience and the industry as a whole (Richards 2011). Broadly, "creative industries" may be defined as "those industries that are based on individual creativity, skill and talent with the potential to create wealth and jobs through developing intellectual property" (Mandic, Petric, and Pivcevic 2017, 336). Creativity itself has been conceptualized using four approaches: creative people, creative products, creative process, and creative press (or environment), these are known as the four *P*s of creativity (Richards 2011, 1226). Each of these creative elements should be considered in film and tourism collaboration for their ability to enhance creative tourism. Film, as a creative industry, has the potential to add creative dimensions (people, products, and process) to a destination (Mandic, Petric, and Pivcevic 2017) that can enhance the potential of creative tourism by forming a creative environment that can then provide an advantage for tourism developers and promoters. As Tomaz (2021) articulates in chapter 2 of this volume "tourism has become an integral part of these development strategies, with local providers identifying and exploiting distinctive endogenous cultural assets." (p. 61) This chapter will consider the four Ps of creativity, and especially their ability to provide insight into the role film and tourism government departments and industry groups can play in creative tourism development.

Film and Tourism Collaboration Case Studies

To better understand how film and tourism collaborations lead to creative tourism, an examination of existing literature and original case-study data is discussed in this chapter. The case-study method and the systematic investigation it enables provides a holistic description of a phenomenon (Lune and Berg 2016). A collective cross-case study was undertaken and involved a number of instrumental cases to help better understand the context of film and tourism collaboration (Stake 2000). Instrumental cases that could provide meaningful insight into film and tourism collaborations were included in the data collection. As an exploratory case-study design, this study did not include a theoretical framework in advance of the data collection. Instead, theory was reviewed at the conclusion of the study when considering implications. To provide a comprehensive understanding of film and tourism collaboration, both questionnaires and interviews with Canadian and American film and tourism professionals were completed. Jensen and Rodgers (2001) suggest that by combining qualitative and quantitative data within a comparative case study, researchers are able to identify contrasts and connections between cases. Questionnaires were used to both collect data and identify cases for further study. Data was collected online by emailing 223 Canadian and American destination marketing organization (DMO) and film commission offices based on a search of all DMO and film offices in each state and province. In total, 18 organizations contacted by email responded to the questionnaire. Questionnaire items were intended to be used as descriptive case-study data, not survey data to represent a population, and therefore the low response rate of 8 per cent was not considered a limitation. Film and tourism industry professionals who responded to the questionnaire were asked if they would be willing to participate in an in-depth interview. Five film and tourism industry representatives who responded to the questionnaire were interviewed to capture specific examples and descriptions of collaborative opportunities and outcomes. These industry representatives were senior officials/staff in their respective regions' film or tourism offices.

The questionnaire asked industry professionals about opportunities and the feasibility of collaborating with the other industry with regards to marketing and communications, advocacy, and sharing resources. Each of these categories contained many of the items drawn from the existing literature. The questionnaire measured both how common and how feasible industry professionals felt collaboration with the other industry was. Items were measured using a five-point Likert-type scale. The interviews were semi-structured and asked film and tourism industry professionals to describe and discuss the nature of collaborations across industries and to identify ways in which these collaborations were facilitated or hindered.

Interview data collected expanded on findings from the questionnaire and revealed opportunities for film and tourism collaboration, the potential outcomes of such collaboration, and factors that would affect it; these are discussed in turn. Finally, planning strategies for enhancing benefits to small communities through film and tourism collaboration are considered.

Opportunities for Film and Tourism Collaboration and the Implications for Small Communities

Findings from the literature review, in-depth interviews, and questionnaires revealed three broad areas for collaboration: marketing and communication, lobbying and advocacy, and resource sharing and development. These areas contain many opportunities for collaboration, as outlined in table 9.1 below.

Table 9.1.

MARKETING/PROMOTION/COMMUNICATION

Receive requests from a DMO (or other tourism body) for images/footage from film/TV/videos produced in your region for use in tourism promotion

Receive images/video produced by the DMO for use in film-location promotion

Share information about film/TV/video awards from the film/TV/video industry with the tourism industry

Share timeline about film/TV/video's release with the tourism industry

Receive timeline on tourism-attraction development from the tourism industry

Share information about release dates of film/TV/video to facilitate tourism marketing

Receive communication from the tourism industry about tourism product development to facilitate film production

Produce location/attraction ads to be featured in key tourism markets

Work with the tourism industry to produce destination ads to be featured in film/TV/video industry key markets

Work with the tourism industry to have destination credits in film/TV/video

Work with the tourism industry to undertake research on key tourism industry markets

Work with the tourism industry to undertake research on key film/TV/video industry markets

Share information on key film/TV/video industry markets with the tourism industry

Receive information on key tourism industry markets from the tourism industry

Jointly promote film/TV/video production with the tourism industry

Jointly promote the destination with the tourism industry

Work with the tourism industry to advertise the destination through media

Work with the tourism industry to advertise a film/TV/video through media

Encourage media coverage of film/TV/video mentions of the destination

Encourage media coverage of the tourism destination mentions of film/TV/videos

Invite travel media to film/TV/video location

Invite travel media to special release of the film/TV/video

ADVOCACY

Work with the tourism industry to advocate for production incentives or subsidies for the film/TV/video industry

Work with the tourism industry to obtain free government services for film/TV/video (access to buildings and parks, parking, permits, police protection, etc.)

Work with the tourism industry to advocate for the development of infrastructure for film/TV/video

Work with the tourism industry to advocate for incentives or subsidies for the tourism industry

Table 9.1. (*continued*)

ADVOCACY
Work with the tourism industry to obtain free government services for tourism (access to buildings and parks, parking, permits, police protection, etc.)
Work with the tourism industry to advocate for the development of infrastructure for tourism
Work with the tourism industry in advocating for the protection of resources (urban and rural landscapes, natural parks, heritage buildings, other)
Work with the tourism industry to inform the government of market changes that impact the competitiveness of the local film industry
Work with the tourism industry to inform the government of market changes that impact the competitiveness of the local tourism industry
RESOURCE SHARING
Develop infrastructure in collaboration with the tourism industry
Provide or facilitate the provision of technical services (e.g., skilled labour) to tourism industry
Provide or facilitate the provision of support services/information (e.g., location services) to the tourism industry
Share or facilitate the sharing of infrastructure (e.g., buildings) with the tourism industry
Share or facilitate the sharing of expertise (e.g., location information) with the tourism industry

Note: The wording presented in this table was intended for the film industry respondents. The items were reworded for the tourism industry. Scale anchors to measure how common activities were included "never," "rarely," "sometimes," "often," "always," and "feasible." These were measured with the anchors "not at all," "slightly," "somewhat," "moderately," and "extremely."

Marketing and Communication Activities

In total, twenty-two marketing and communication activities that tourism and film organizations can undertake together were described based on the literature review, interviews, and questionnaire.

Marketing activities were divided into sharing (information and marketing material), producing (marketing material, advertisements, and research), and promoting (to the public, to media, and to the industries).

The questionnaire revealed that marketing and communication activities were considered the most feasible area of collaboration (m = 2.02). Survey participants suggest that this is because of the need for both industries to undertake profile-raising and image-enhancing activities to achieve their mandates, and that these activities are relatively easy to undertake in collaboration.

Within this area of collaboration, the most feasible opportunity for collaboration was sharing and receiving images/footage (m = 2.66). The least feasible was ensuring tax credits for featuring the destination as itself (m = 1.29). With regards to how common various activities were, sharing and receiving images and footage was the most common collaboration with regards to marketing and communication, with 14 out of the 18 respondents indicating this is an activity they've participated in. The least common activity undertaken in collaboration was researching film/tourism key markets, with only 7 of the 18 respondents indicating this was an activity undertaken in collaboration.

An example from Canada's Northwest Territories (NWT) demonstrates how advertising can be shared. The tourism office purchased a commercial spot to coincide with the airing of episodes of the CBC television show *Arctic Air*. The NWT also secured ad space on the show's CBC streaming website. This contributed to promoting the NWT as a destination across the country.

Another example from the NWT highlights the role of collaboration. In 2014 the NWT Department of Industry, Tourism and Investment launched a road visitor survey that included items related to television- or film-induced tourism. This demonstrates how collaboration can lead to data being used by both the tourism and film industries. In this case the survey undertaken by the NWT government tourism office provided insight into the value film and tourism offer each other for drawing visitors to the destination.

One film official described the value of destination-familiarization tours featuring film locations. They noted that when the tourism industry hosts media professionals, media is encouraged to visit past and potential film locations, which contributes to raising the destination's profile as a feasible film location. For example, the New Mexico Film Office and the tourism department collaborated to produce film-tourism maps of the region to encourage visitation to film locations within the state (New Mexico Film Office n.d.). Furthermore, Visit Albuquerque used celebrity endorsements to highlight elements of the city that appeal to visitors—for example, the following statement from actor Bryan Cranston: "From the

Sandias glowing at sunset to heading out to To'hajiilee, there's pure magic that the city lends to the show" (quoted in Visit Albuquerque n.d.).

Overall, the in-depth interviews showed that marketing and communication collaborations, while often serendipitous, are on occasion the result of deliberate planning. When sharing information between film and tourism industries, those interviewed stated that information is often exchanged through newsletters and listserves that include both industry groups, and so members of both sectors passively engage with each other's organizations. There are examples in which a region's film office actively shares information with the tourism office, but this is less common. Whether information is purposefully exchanged or not seems to depend on the formality of the relationship between the film and tourism offices. The more established the overall relationship, the more often they actively share information. It is possible that through planning, the film and tourism industries can develop strategies to remain actively engaged with one another. At the local level, the creative activities of the film industry can play a critical role in the development, promotion, and distribution of tourism products (Mandic, Petric, and Pivcevic 2017). By collaborating on marketing and communication activities, destination promotion contributes to the enhancement of the destination's image and the potential development of film tourism. Richards and Wilson (2006) note that the stories about place offered through film can enhance perceptions of a destination. Film has the ability to increase tourists' awareness of a location while presenting its characteristics in a visual and emotional format (Cardoso et al. 2017). Furthermore, the perception of a place as creative enhances its overall attractiveness (Richards 2011). Therefore, when the film and tourism industries collaborate on marketing efforts, new creative products (film-tourism product development) are promoted and a creative environment (through the image of the destination as a worthy location for filming) is offered to the public for consideration.

Advocacy and Lobbying

In total, nine opportunities for collaboration within the realm of film/tourism advocacy and lobbying were revealed through the literature review, interviews, and questionnaire. Advocacy is not an option for some film or tourism organizations as their mandate and relation to government doesn't allow for this.

Based on findings from the literature review, questionnaire, and interviews, advocacy activities were divided into two broad categories: (1) obtaining resources that benefit film and tourism (incentives, services, infrastructure, resource protection), and (2) providing information about the value of film and tourism (information, research).

The questionnaire showed that lobbying and advocacy activities were not considered very feasible (m = 1.39). The most feasible area of collaboration was considered working with the other industry to advocate for incentives (m=1.79). This was also considered the least common area of collaboration within advocacy and lobbying (33 per cent). Respondents shared that the most common opportunity for collaboration is informing government of changes that impact industry competitiveness (55 per cent).

An example of advocacy to obtain resources was offered by a representative from a film office when they noted that they would like to work with their tourism industry counterparts to advocate for additional funds from government that could benefit both industries. They described using the additional funding for promoting the destination/location. Interviewees noted that government funding for their activities was insufficient and at times unreliable. Partnering was considered an effective way to show greater value to government for their investment.

The need for advocacy that would provide information on the value of film and tourism to government was highlighted by two representatives, one from a state film office and the other from a state tourism office. They felt that because of the complimentary nature of film and tourism, it was important that the two industries work together to raise the profile of the region. One interviewee envisioned tourism and film working in conjunction to promote each other in government forums, and one state representative noted the possibility of testifying in front of legislative committees on issues that are of mutual interest to film and tourism (such as the preservation of historic districts).

The in-depth interviews described why advocacy is challenging to undertake in collaboration. The main reason these activities are not always considered feasible is because some organizations are not in a position to advocate. Lobbying is formally restricted for some organizations, preventing them from engaging in a variety of activities. However in most situations, advocacy activities such as informing and educating are considered appropriate. Competition also affects advocacy collaboration. One film office interviewee stated that in some states, members of the film and tourism industries feel they are competing for scarce resources and as such don't want to advocate together. Another state tourism representative noted the capacity of these industries to bring different groups together (through such things as attractions, accommodations, transportation, etc.), and by including film officials in meetings with other industry groups, they saw the potential for more diverse industry networks to form and to advocate across industries.

By working together to advocate and lobby, both industries can receive benefits that in turn contribute to enhancing film and tourism opportunities. When the local film industry is strong, the community's creative environment is enhanced, which in turn contributes to the image of a creative destination that can be leveraged by both industries.

Resource Sharing and Development

Many examples exist in the literature of resources being shared across industries (Hudson 2011). In some cases tourism and film collaborate by offering cash incentives and logistical support to productions that feature the destination as itself (Hudson 2011). This is useful if the community would like to use the film industry to develop film tourism or enhance the destination image.

The examples of resource sharing fit within three categories: providing a resource to the other industry, sharing a resource, or jointly developing resources.

The questionnaire revealed that resource sharing and development was considered somewhat feasible (m = 1.92) by tourism- and film-industry professionals who responded. The most feasible area of collaboration was the sharing of expertise (m = 2.75). The least feasible collaborative opportunity was the collaborative development of infrastructure (m = 1.08). The most common sector for collaboration in this area was location services (78 per cent), whereas the least common was direct sponsorship across industries (39 per cent).

Interviewees noted that resource sharing is considered a valuable collaborative opportunity, but that resources can only be successfully shared when both parties are committed to working together. This is another example of the need for planning that will be discussed later in the chapter.

A tourism representative highlighted a unique example in which resources were provided to the film industry. The film industry in the region regularly relied on the tourism industry to help with finding and providing access to remote/extreme locations. Through opportunities like this, local tourism guides were employed by a film producer to escort crew members to locations, thus offering new employment opportunities for residents. There are also many examples of films leaving resources (renovated restaurants, piers, properties) behind that benefit the community and tourism either directly or indirectly (Lundberg, Ziakis, and Morgan, 2017).

An excellent example of sharing resources came from a tourism and film office that had recently been merged. The representative interviewed stated that now that they were formally connected they were better able to use seemingly disparate information to inform decisions. In this case, the tourism office was now privy to information about productions in the early stages of development. This information was used to develop promotional strategies to coincide with the release of a film.

While joint resource development is less common, an example was nonetheless given in which the New Mexico Tourism Department produced a local television show called *New Mexico True*, employing local crew members (New Mexico Tourism Department, n.d.). This helped to support those individuals on an ongoing basis and contributed to the show's ability to feature local destinations. In addition, the New Mexico Tourism Department and the New Mexico Film Foundation jointly funded the documentary series *Life in New Mexico* to develop skilled labour in the film industry and promote tourism in the state using imagery highlighting life in New Mexico.

Resource sharing can contribute to the community's creative offerings. Examples provided above highlight new creative ventures that result from the industries sharing resources (including labour). Through the development of skilled film labour, creative people are attracted to a community, and they in turn contribute to the creative environment. Film

Figure 9.1. Opportunities for Film and Tourism Collaboration

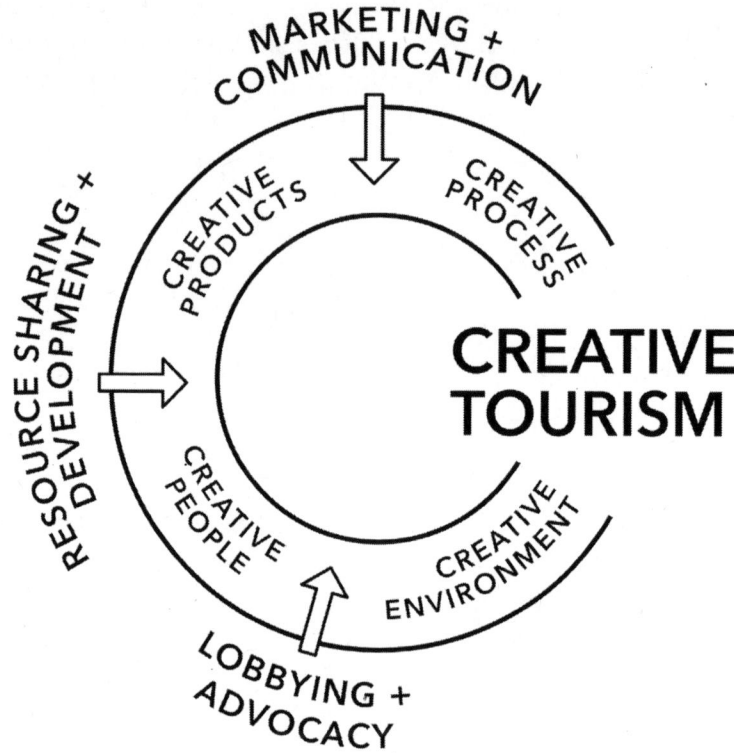

projects may also result in the development of resources that, when left behind, can result in new elements that enhance the community (unrelated to film itself).

By examining opportunities for collaboration, this research revealed that marketing and communications, advocacy and lobbying, and resource sharing and development are all possible directions that should be considered. These activities' contributions to creative tourism are presented in Figure 9.1. Through marketing and communications, the destination's image can be enhanced to benefit both industries. This image is connected to the creative process. Through the creative activity of filmmaking, a destination's unique "story" is created and consumed as a creative product. Its identity, personality, and image are developed and portrayed through the film, conveying the brand's identity to the viewer (Cardoso et al. 2017). The creative product evolves out of the experience that is produced. By sharing resources, attractions are developed for consumption. Finally, through lobbying and advocacy a stronger film and tourism industry is formed, attracting creative people who can contribute to a creative environment.

Outcomes of Film and Tourism Collaboration for Small Communities

In this section, both the benefits and drawbacks of film and tourism collaboration will be examined and discussed in regards to creative tourism development.

To date, much of the research on the implications of film tourism have been single case studies (Heitmann 2010), looking at specific locations such as Dubrovnik (Tkalec, Zilic, and Recher 2017), Kazakhstan (Pratt 2015), or South Korea (Kim et al. 2007). Outcomes related to film and tourism development examined in past research have typically focused on film-tourism products and are not specifically attributable to film tourism but rather to tourism impacts generally (Heitmann 2010). The existing scholarship shows that employment opportunities, tourism infrastructure, diversification of product, host community interaction, and cultural exchange, as well as conflict, commodification, loss of authenticity, and cultural and natural depreciation, have all been discussed in relation to film tourism (Heitmann 2010; Cardoso et al. 2017; Mandic, Petric, and Pivcevic 2017). Here, the outcomes of collaboration between these industries will be discussed to help better understand how this collaboration can be leveraged for creative tourism development that benefits communities rather than resulting in negative implications.

The in-depth interviews and questionnaires undertaken to better understand collaboration between film and tourism identify outcomes of collaboration that fall within four broad categories: industry and economic development; destination/location image; community development; and film-tourism product development.

Industry and Economic Development

Through collaboration, film and tourism industries grow and diversify. The film industry can create a "cascading effect throughout the local economy," affecting many sectors including tourism, although the exact nature and extent of this impact can be difficult to measure due to broad scope of possible direct, indirect, and induced effects (Mandic, Petric, and Pivcevic 2017, 339). Further complicating this assessment is the amount of leakage that may occur when using non-national film companies and crew. The film industry professionals interviewed described the benefits of the tourism industry employing local crew on tourism-related productions. These experiences allowed crew to hone skills and obtain work during slow periods (as production is sporadic in many regions). In some instances, local crew members were employed to produce advertisements and other footage for tourism promotion. The tourism industry benefited from the hiring of knowledgeable crew who can film under extreme weather conditions and are familiar with locations. Tourism workers employed on film productions also benefited both industries. Respondents noted that tourism professionals had knowledge of the local area and communities that visiting film productions benefited from. Ultimately, through film and tourism collaboration, creative members of the community are supported and may be better able to sustain themselves in the long-term, which draws creative people to the community.

Additionally, the research shows that by working together, the film and tourism industries were able to maximize their purchasing power. For example, by facilitating the joint purchase of stock footage of the region, one film office was able to acquire footage that could be used by both industries. This contributes to the development of creative products through image enhancement by including high-quality imagery in both creative media (film, television, and digital) and tourism-promotional material.

Destination/Location Image

The perceived image of a given destination is a crucial element in its success (Chen 2018). This is also true regarding the image of film and the industry. The profile-raising capability of each industry is mutually advantageous and has the potential to benefit communities, even though the films are rarely created with the intent of attracting tourists to the location (Juskelyte 2016). Increased exposure enhances both film's and tourism's ability to attract creative people to a region and thereby contribute to a creative environment. In one state, tourism and film were placed within the same department and a department representative noted film's ability to increase the destination image and for tourism to enhance the attractiveness of locations for filming. Worldwide, there are many examples of these industries working together to promote locations as destinations (Roberts 2012). Visit Britain was one of the first to produce movie maps, and this remains a popular product for guiding visitors. These tools are important for marketing and managing destinations (Beeton 2005).

Even productions featuring stories that initially seem unlikely to benefit a location can contribute to enhancing its profile and its ability to draw tourists. The highly acclaimed television series *Breaking Bad*, for example, features controversial content that seems unlikely to enhance the image of Albuquerque, New Mexico. However, the profile-raising effects of the series is unquestionable, and tourism to Albuquerque increased as a result. Local businesses have benefited from increased visitation and tourist spending. The series *Chernobyl* and *Narcos* both offer examples of challenging or upsetting content driving interest in a destination. Indeed, despite the tragic nature of the events of the Chernobyl nuclear disaster, the exclusion zone, which remains contaminated, has nonetheless attracted substantial interest from visitors (Kolirin and Guy 2019). The violence inflicted by narco-terrorist Pablo Escobar devastated many communities throughout Columbia, and yet tourism to sites related to the terrorist is growing (Tanzi 2017).

Not only do film and television portrayals of a destination raise interest in it, but endorsements offered by high-profile celebrities in the film industry can also contribute to enhancing this image. Celebrities can affect the perception of a location in a variety of ways, including by raising awareness of the destination, projecting their own image onto the place, and allowing their fans to live vicariously through them (Chen 2018). Additionally, this may be achieved by associating creative people with the destination, thereby creating an image of a creative place populated by creative people. The film *Dear John*, for example, helped contribute to film tourism to locations in South Carolina. After filming in the state,

Channing Tatum commented that "I love South Carolina. . . . I'm from the South, so I have an addiction with it. The food, the people, the lifestyle. It's just so charming." Amanda Seyfried, lead actress in the film, was quoted as saying, "the South is a whole different thing altogether. It's romantic. It's just so beautiful. It brings something to a film you can't get anywhere else in the world" (SCIWAY n.d.). Each of these endorsements connect creative people with a place and also highlight attractive elements of the destination for visitors to consider (Chen 2018).

Roberts (2012), though, argues that through film and tourism collaboration, place can be further commoditized, which can have the effect of undermining the destination's unique identity. This is problematic, as differentiation is an important aspect of a destination's image (Blain, Levy, and Ritchie 2005). Roberts (2012) does identify an opportunity to move beyond simply publishing lists of film places appropriate for the tourist gaze. Instead, he suggests integrating more meaningful information (involving oral histories and interviews) to provide a meaningful context for a location.

Community Development

Existing literature provides extensive evidence that the film and tourism industries can positively and negatively impact the communities in which they operate (Kim et al. 2007; Heitmann 2010; Beeton 2016; Croy 2010; Lundberg, Ziakis, and Morgan 2018). The precise nature of this impact depends on the scale and characteristics of the production or tourism development, and the characteristics of the community (Wray and Croy 2015). When film and tourism organizations collaborate, there are opportunities to reduce the negative implications and maximize the benefits of both industries to the communities they depend upon. Previous research features evidence of residents' dissatisfaction with the tourism and film industries because of issues like cultural commodification, crowding, and loss of privacy (Beeton 2008; Roberts 2012). There are many examples of destinations gaining in popularity because of film or television, yet local residents may not want increased tourism, as in Dubrovnik (Tkalec, Zilic, and Recher 2017), which can be problematic (Wray and Croy 2015). Beeton (2010) recognized the need for the community to drive tourism development in order to ensure that the resulting products meet the needs identified by the community. With thoughtful planning and assistance, these problems can be addressed and the benefits of both tourism and film maximized (Beeton 2008).

Representatives from regions across Canada and the United States provide examples of film legacies that benefited communities through collaboration with the tourist industry. For instance, there are examples of film productions donating set items and clothing to local groups throughout both countries; such was the case with the production of *Army Wives* in South Carolina, which donated goods to the local community. There are also many examples of film productions repairing or even building infrastructure within the community. The community's access to these resources and how they are used determines the long-term impacts of these legacies. For example, the film *Dear John* undertook a restoration of the fishing village in which it was shot in 2009; the community continues to benefit from these improvements and the island location remains a tourist destination today. Many of the industry professionals interviewed noted that a strong relationship can be forged between the production and the community in which it is shooting. These relationships can work to enhance the legacy of such projects; however, this doesn't occur serendipitously and requires conscious effort.

Industry professionals interviewed noted that larger studios/production companies often employ community-relations professionals who find opportunities to contribute to the community as part of a production's corporate-responsibility mandate. Unfortunately, smaller productions, with limited resources, are not always able to benefit the community. Interviewees noted that the local film office or tourism department can help provide benefits by facilitating relationships between productions and the community.

Film Tourism and Creative Product Development

While both film and tourism can contribute to the local community broadly, film tourism is a unique opportunity for a destination to leverage film and television productions in order to enhance its creative tourism product offerings. Film professionals interviewed noted a tendency for film officers to focus their attention on industry needs rather than their own community. The reason given was a lack of resources and a need to stay focused on their primary mandate. Film-tourism opportunities are not often considered; instead the focus is on attracting productions for the benefit of the local economy, with relatively little attention paid to the long-term spinoff benefits (Hudson 2011). Beeton (2008) notes that most regions recognize the financial inflow from the initial film production without giving much thought to the long-term possibilities that film tourism offers.

In order to develop successful film tourism, proactive collaboration between film, tourism, and community groups during the early stages of production is ideal (Lundberg, Ziakis, and Morgan 2017). Many elements of product development were identified through the interviews, questionnaire, and literature reviewed. Some key areas include leveraging promotional media both before and during production, negotiating end credits that highlight locations, and obtaining quotes from high-profile film cast and crew during and after production.

Film-tourism products include more than mere location visits; tangible product development is also a possibility. For example, *Ice Pilots NWT* features Buffalo Airways. Buffalo Airways was able to benefit not only from the popularity of the series but also from the resulting sale of Buffalo Airwear.

Both film and tourism rely on the communities in which they exist. When community members reap the benefits of film-induced tourism, they are more likely to offer valued experiences to their guests. Alternatively, if hosts do not appreciate film tourism they will not create a positive experience (Beeton 2008). This suggests a need for planning that includes the community. This will be discussed at more length later in this chapter.

Factors Affecting Film and Tourism Collaboration

Many examples of collaboration between the film and tourism industries have already been provided. Interviews and questionnaire respondents reveal three main factors that appear to affect opportunities for collaboration. Representatives of the film and tourism industries highlight the roles of personal relationships, a collaboration champion, and organizational structure as the main drivers of the success of this partnership.

Personal Relationships

All five interviewees discussed their personal relationships with their counterparts in the other industry. They felt strongly that the relationships between individuals make collaboration more or less likely. Four of the interviewees talked about how their desire to work with their counterparts in the other industry made collaboration possible. One interviewee from a state film office described how their local tourism department was not interested in working with members of the film industry, which made collaboration challenging. When asked about community involvement, the interviewees provided similar answers. Existing personal relationships are key to ensuring the film and tourism industries benefit the local community.

In chapter 10 of this volume, Prince, Petridou, and Ioannides highlighted the value of networks of artists embedded in place. Within the context of film and tourism collaboration, the value of the connection between these industries, spawned by their reliance on local landmarks, locations, and destinations should not be overlooked. However, there are challenges in the development of relationships, and issues of power especially must be considered.

Beeton (2016) discusses the role of power in such collaboration, stating that, while film might hold creative control, tourism often has power over location access. When strong positive relationships exist across these industries, collaboration may become possible despite differences in power. Beeton (2016) also recognizes the important role of the destination-marketing organization and film commission as mediators between industries helping to distribute power.

Community Champion

Most film and tourism representatives interviewed described a champion who worked toward strong relationships and successful collaborations between film and tourism. These champions emerged from various places. At times, it was the head of either the film commission or tourism marketing department, but government officials also played an important role. In one state, it was the governor who brought film, tourism, and the local community together to enhance the benefits to the state.

Organizational Structure

The structure of existing government departments and community organizations is a factor in successful collaboration. The literature review suggests that when tourism and film are brought together in the same government department, they collaborate at earlier stages to leverage film for tourism (Hudson 2011).

One of the interviewees described a case in which film and tourism were in the process of being brought together under a new organizational structure. The head of this department felt that because of their integration, the industries would work together to enhance the profile of the state, with both acting as "stewards of the destination image." While this speaks specifically to the perspective of tourism, this executive felt that by sharing a range of resources, each industry would benefit from new knowledge and approaches. Furthermore, depending on the government structure within the state or province, the film and tourism industries' mandates take on different perspectives. Based on interviews with industry professionals, the perspective taken can influence the industries' level of community engagement. For example, depending on whether the local film and tourism agencies report to economic development departments or culture departments influences their ability to effectively collaborate. A film and a tourism representative in one state highlighted the fact that there is little direct collaboration between the sectors because of their current structures; film is housed within the economic development branch of the government, while the state tourism agency is an arm's-length organization. In this case, the personal relationships between the film and tourism offices fostered a great deal of collaboration, but the official structure was a constant barrier that the staff had to negotiate.

An approach to formalizing the relationship between film and tourism industry groups was presented by the one film office. In this territory, film and tourism are not in the same department, but the film office identified tourism as a stakeholder in their film strategy report. Specific action items are contained within the strategy, thus ensuring tourism and film organizations work together for mutual benefit. Another film industry professional noted that without a formal strategy, film production tends to be reactive to the current needs of the film production or industry as a whole. By working together in a strategic and proactive manner, greater benefits can be achieved. The same is true for community engagement. When the community is considered a key stakeholder, it is has a great say in how film and tourism are leveraged to its benefit (Wray and Croy 2015).

Even when the responsibility for film and tourism is held by the same department, it is possible for them to remain at odds. In one state, a senior staff member in the film office noted that tourism is "entrenched in replicating the past" and is hesitant to engage with film. The film office representative indicated that they felt that the local tourism office didn't want film taking on any of the work the tourism office currently undertakes. In a case like this, a strong relationship between individuals or a collaboration champion could facilitate co-operation.

There is limited research exploring the challenges to collaboration between the film industry and tourism. Film tourism research shows that a destination's tourism potential is not always achieved after production (Beeton 2008). Within the collaboration research more broadly, "we versus me" thinking, ownership, and turf wars are identified as issues preventing collaboration (Lewis, Isbell, Koschmann 2010). In our study, the factors preventing collaboration were explicitly explored. These include lack of will, time, money, return on investment, control, and collaboration not being within the scope of the organization's strategic direction.

Planning Strategies to Enhance Benefits to Small Communities through Film and Tourism Collaboration

Opportunities for film and tourism industry collaboration are presented throughout this chapter. While factors affecting collaboration are noted, there is still a need to examine how creative tourism can be explicitly developed from film and tourism collaboration. Past research reveals that destination image enhancement and film tourism development are not always leveraged during post-production, reducing the value of film production to the destination (Hudson 2011), especially when the destination is unprepared for the opportunity (Volo and Irimias 2016). Beeton (2016) notes many benefits, beyond enhancing the destination's image, that a film production offers a community. Specifically, Beeton suggests entrepreneurial business. While most commonly these opportunities involve guided tours, it is noted that film-themed "products, places, souvenirs and guidebooks" (108) can be developed for the market. By working together, film and tourism can find ways to benefit a place and help both industries prosper.

To accomplish this, tourism-planning and -development concepts will be examined in relation to film and tourism collaboration. Tourism can begin with a single attraction (such as a film location). In the case of film tourism, a small community can quickly be overwhelmed by tourists, which changes the destination dynamic (Reid, Mair, and George 2004; Tkalec, Zilic, and Recher 2017). It is believed that through appropriate planning, sustainable creative tourism can be leveraged from film and tourism collaboration.

Within the existing film tourism literature, marketing, product development, and image creation/enhancement have generally been the focus of research, in addition to the motivations and perspectives of film tourists (Cardoso et al. 2017); limited research has been conducted into planning as it relates to film tourism (Heitmann 2010). However, planning within tourism more broadly is a well-researched phenomenon (Reid, Mair, and George 2004). Within film tourism, various approaches have been suggested for strategic planning (Beeton 2016; Heitmann 2010). Mandic, Petric, and Pivcevic (2017) suggest four aspects to encouraging film tourism: proactively encouraging filmmakers and producers to come to the region; developing publicity for the film and location; marketing plans for post-production; and peripheral marketing to fully maximize the potential. However, these strategies ignore a crucial element in the success of the process, the stakeholders. Stakeholder theory has commonly been applied to conceptualize tourism planning (Heitmann 2010), and power has been examined as a critical issue (Beeton 2016) that should be considered.

Stakeholder theory is well-established and informs community tourism planning. Stakeholders are those who have an interest in the activities of an organization. By understanding the different groups' interests, their perspectives can be included in plans. Stakeholder theory highlights the importance of stakeholder management and relationships (Heitmann 2010). In the context of collaboration between the film and tourism industries, there are many obvious stakeholders that should be involved in any planning for tourism development. Destination management organizations (DMOs), the local community, tourists, tourism businesses, and the film industry should all be included, according to Heitmann (2010). To expand Heitmann's recommendation, the following industry groups should also be part of the planning: attractions; food and beverage; accommodations; transportation; tourism services; and the travel trade. Heitmann does not elaborate upon the film industry groups that should be involved. Based on the interviews undertaken to better understand collaboration between the film and tourism industries, it seems the following film industry stakeholders should be included: the destination film liaison office; the film industry association; the film commissioner's office; local producers; and directors as well as location managers. As Heitmann notes, a film is not likely to be involved in the early stages of tourism development planning. Neither are they likely to be involved in planning on an ongoing basis because of their primary focus on the film itself. The film commission office and local industry association have the potential to engage early, and to remain engaged in the planning process; as a result, they should take on a primary role in planning alongside the DMO, which is also engaged with the community.

Based on the findings from the interviews described above, it is clear that stakeholder management needs to encompass the power between the various individuals and groups involved in tourism and film industry collaboration. Community-based tourism development research has demonstrated that stakeholders with powerful economic interests often dominate decisions around tourism development, while less-powerful community groups receive less attention (Richards and Hall 2003). Stakeholders' roles and interest in film tourism planning varies, and thus the level of involvement and participation in the planning process differs across stakeholders (Wray and Croy 2015). On occasion, industries may be concerned about a loss of control when collaborating, which could impede possible opportunities to work together (Hudson 2011). Given the need to start planning early and to maintain that process over the long term, the ideal scenario should involve a tourism industry champion working in collaboration with a film industry champion to formally engaging in planning. An issue for many small communities that serve as film locations is that it is not common for any of the film industry professionals to reside in the community or maintain ongoing connections with its members.

Small communities within a few hours' drive of a larger centre may act as a location for only a few key scenes. This is where the film commission office for a province, territory, or state can be particularly critical to attempts to leverage tourism to benefit the community.

The value of local community involvement in tourism planning is well-known (Beeton 2005), yet within the film context we have seen that the needs of the community are often not considered. Beeton (2008) notes that the time between initial filming of a production, its release to the general public, and film tourism can be long (often years), which can be a challenge for communities who want to develop film tourism or use film production to enhance their image. Yet while it may be challenging to cultivate immediate benefits from film productions, this delay offers an opportunity for the community to plan. Changes in technology and increasing competition within the film industry may be shortening the times between production and film release, which in turn could compress the amount of time available to a community to plan. Again, a champion may be needed to bring stakeholders together to take advantage of this compressed time frame.

Figure 9.2 highlights the role of and the relationship between the various stakeholders described above. By involving these groups in the development of creative tourism based on film and tourism collaboration, successful film tourism and related opportunities are possible. How these groups should engage in this process can be informed by Timothy and Tosun's (2003) planning principles, which are summarized as "PIC" (participatory, incremental, cooperation/collaboration). This framework for sustainable planning involves participatory development, incremental growth, and collaborative efforts. Participatory development includes involving stakeholders, as outlined above, in the development process. The community is believed to be empowered when all stakeholders are encouraged to work together to identify goals, programs, and projects. The PIC principles suggest that once plans are in place, constant progress is monitored to ensure the various goals outlined are achieved and that principles of sustainability, such as cultural and ecological integrity, holistic development, and balance and efficiency, are observed. Finally, collaboration between varying groups is necessary, including between government departments, different levels of administration, and public and private groups (Timothy 2011).

Figure 9.2. Film and Tourism: Roles and Relationships

Once groups are brought together to develop creative tourism based on film and tourism collaboration, it is necessary to identify the relevant opportunities for the specific community. The creative products, processes, people, and environment should all be considered by the stakeholders. A method that may be useful for identifying opportunities is culture mapping (Creative Cities Network of Canada 2010). Culture mapping takes many forms, but at its core is the identification of cultural resources by stakeholders that can then be placed on a map based on dimensions. In this way, culture mapping helps communities to understand and share their culture in a geographic space. Through this collaborative process, cultural planning informed by varying perspectives is possible. This is closely aligned with the PIC model. As Hall (2012) states, spatial analysis is useful for tourism planning and spatial modelling can offer insight into the opportunities for creative tourism. In describing film tourism maps more broadly, however, Roberts (2012) suggests that "visual commodification of the location site in the form of the movie map and other film-related tourism media does little to put the 'real' city on the map" (194). To avoid this pitfall, the culture-mapping process should ensure local perspectives are represented in the map.

Through culture mapping, creative clusters are identified. This allows stakeholders to better understand how resources are related. For example, all creative resources related to both film and tourism placed on a geographic map of a community will reveal opportunities. Where there are overlaps and a will from the community to develop tourism and share cultural resources, the community should consider developing creative tourism opportunities. By bringing creative elements together, stakeholders form creative clusters that are thought to result in mutually beneficial "spinoffs." However, Richards and Wilson (2006) warn that if creative clusters become stagnant, the benefits of grouping creative elements together may not be realized. The PIC model suggests that incremental growth can alleviate this risk (Timothy and Tosun 2003).

While many well-known and successful examples of film tourism can be seen throughout the world, there are many locations that never benefit from the film productions in their environs or evolve into attractions, despite their potential. As Richards (2011) notes, destinations' "organizational capacities allow some regions to make better use of their inherited and created assets to make themselves attractive to tourists" (1231). Of those locations that do become attractions, the outcomes are not always positive for the local community. Developing a strategy for a region by working collaboratively across industries and with the community will help ensure that opportunities are not missed and that unique creative tourism products are developed that meet the needs of the community. Furthermore, through planning, the film and tourism industries can develop strategies to remain actively engaged with one another in order to benefit the community

Conclusions

Research findings reveal that both unlikely destinations and improbable scripts can lead to a surge in interest in a city or region by presenting a unique narrative tied to the destination. Currently, the vast majority of communities will only see short-term benefits from films shot on location in their neighbourhoods. Without a developed film tourism strategy (and its accompanying resources), it may take years of post-production to develop the final product, and this may result in a missed tourism opportunity. Currently, many landmarks, landscapes, and locations used in filming develop into tourism attractions organically and sporadically but not always with support from the stakeholders (Wray and Croy 2015).

With a strategy in place to harness creative, people, products, processes, and press, opportunities can be developed to maximize the benefits of film for small communities. Missed opportunities for film tourism also exist when the material legacy left behind by a film production is not used to further develop tourism attractions and opportunities. Over time, this lack of benefit from filming can create a scenario in which communities begin to resent the imposition of film productions. The fictional film *State & Main* (2000) presents a comedic exploration of the relationship between a production and the town in which it is filmed. In it, the character Walt Price comments that "this is what my people died for . . . the right to make a movie in this town." If sustainable film or tourism industries are to flourish, film and tourism organizations should question how they both engage with and contribute to the community. The time spent filming in a community could result a range of possible creative results if leveraged. By working collaboratively, film and tourism industries can make this a reality.

This chapter presented examples from across Canada and the United States to highlight how collaboration between the film and tourism industries can engage visitors with novel creative tourism products. Film and tourism industry collaboration can lead to sustainable creative tourism, beyond film tourism, in small communities

References

Beeton, Sue. 2005. "The Case Study in Tourism Research: A Multi-method Case Study Approach." In *Tourism Research Methods: Integrating Theory with Practice*, edited by Brent Ritchie, Peter Burns, Catherine Palmer, 7–48. Cambridge, MA: CABI.

———. 2008. "Location, Location, Location: Film Corporations' Social Responsibilities." *Journal of Travel & Tourism Marketing* 24, nos. 2–3: 107–14.

———. 2016. *Film-Induced Tourism*. Bristol, UK: Channel View Publications.

Blain, Carmen, Stuart E. Levy, and J. R. Brent Ritchie. 2005. "Destination Branding: Insights and Practices from Destination Management Organizations." *Journal of Travel Research* 43, no. 4: 328–38.

Bolan, Peter, and Lindsay Williams. 2008. "The Role of Image in Service Promotion: Focusing on the Influence of Film on Consumer Choice within Tourism." *International Journal of Consumer Studies* 32, no. 4: 382–90.

Cardoso, Lucilia, Cristina Estevao, Cristina Fernandes, and Helena Alves. 2017. "Film-induced Tourism: A Systematic Literature Review." *Tourism & Management Studies* 13, no. 3: 23–30.

Carl, Daniela, Sara Kindon, and Karen Smith. 2007. "Tourists' Experiences of Film Locations: New Zealand as 'Middle-Earth.' " *Tourism Geographies* 9, no. 1: 49–63.

Chen, Chien-Yu. 2018. "Influence of Celebrity Involvement on Place Attachment: Role of Destination Image in Film Tourism." *Asia Pacific Journal of Tourism Research* 23, no. 1: 1–14.

Creative Cities Network of Canada. 2010. *Culture Mapping Toolkit: A Partnership between 2010 Legacies Now and Creative City Network of Canada.* https://www.creativecity.ca/publications/ccnc-toolkits.php.

Croy, W. Glen. 2010. "Planning for Film Tourism: Active Destination Image Management." *Tourism and Hospitality Planning & Development* 7, no. 1: 21–30.

Duxbury, Nancy, and Greg Richards, eds. 2019. *A Research Agenda for Creative Tourism.* Cheltenham, UK: Edward Elgar Publishing.

Hahm, Jeeyeon, and Youcheng Wang. 2011. "Film-Induced Tourism as a Vehicle for Destination Marketing: Is It Worth the Efforts?" *Journal of Travel & Tourism Marketing* 28, no. 2: 165–79.

Hall, C. Michael. 2012. "Spatial Analysis: A Critical Tool for Tourism Geographies." In *The Routledge Handbook of Tourism Geographies*, edited by Julie Wilson, 163–73. London: Routledge.

Heitmann, Sine. 2010. "Film Tourism Planning and Development—Questioning the Role of Stakeholders and Sustainability." *Tourism and Hospitality Planning & Development* 7, no. 1: 31–46.

Hudson, Simon. 2011. "Working Together to Leverage Film Tourism: Collaboration Between the Film and Tourism Industries." *Worldwide Hospitality and Tourism Themes* 3, no. 2: 165–72

Hudson, Simon, and Vincent Wing Sun Tung. 2010. " 'Lights, Camera, Action...!' Marketing Film Locations to Hollywood." *Marketing Intelligence & Planning* 28, no. 2: 188–205.

Hudson, Simon, Youcheng Wang, and Sergio Moreno Gil. 2011. "The Influence of a Film on Destination Image and the Desire to Travel: A Cross-Cultural Comparison." *International Journal of Tourism Research* 13, no. 2: 177–90.

Jensen, Jason L., and Rodgers, Robert. 2001. "Cumulating the Intellectual Gold of Case Study Research." *Public Administration Review* 61, no. 2: 235-246.

Juskelyte, Donata. 2016. "Film Induced Tourism: Destination Image Formation and Development." *Regional Formation and Development Studies* 2, no. 19: 54–67.

Kim, Samuel Seongseop, Jerome Agrusa, Heesung Lee, and Kaye Chon. 2007. "Effects of Korean Television Dramas on the Flow of Japanese Tourists." *Tourism Management* 28, no. 5: 1340–53.

Kim, Sangkyun. 2012. "Audience Involvement and Film Tourism Experiences: Emotional Places, Emotional Experiences." *Tourism Management* 33, no. 2: 387–96.

Kolirin, Lianne, and Jack Guy. 2019. "Chernobyl to Become Official Tourist Attraction, Ukraine Says." *CNN Travel*, 11 July. https://www.cnn.com/travel/article/chernobyl-tourist-attraction-intl-scli/index.html.

Lewis, Laurie, Matthew G. Isbell, and Matt Koschmann. 2010. "Collaborative Tensions: Practitioners' Experiences of Interorganizational Relationships." *Communication Monographs* 77, no. 4: 460–79.

Lundberg, Christine, Vassilios Ziakas, and Nigel Morgan. 2018. "Conceptualising On-Screen Tourism Destination Development." *Tourist Studies* 18, no. 1: 83–104.

Lune, Howard, and Bruce Berg. 2016. *Qualitative Research Methods for the Social Sciences*. Harlow, UK: Pearson Higher Ed.

Mandic, Ante, Lidija Petric, and Smiljana Pivcevic. 2017. "Film as a Creative Industry Constituent and Its Impacts on Tourism Development: Evidence from Croatia." *Tourism in Southern and Eastern Europe* 4: 335–48.

New Mexico Film Office. n.d. "Film Tourism." Accessed 19 October 2020. https://nmfilm.com/for-fans/film-tourism/.

New Mexico Tourism Department. n.d. "New Mexico True Television." Accessed 6 April 2018. https://www.newmexico.org/plan/true-television/.

OECD. 2009. *The Impact of Culture on Tourism*. Paris: OECD.

Pratt, Stephen. 2015. "The Borat Effect: Film-Induced Tourism Gone Wrong." *Tourism Economics* 21, no. 5: 977–93.

Reid, Donald G., Heather Mair, and Wanda George. 2004. "Community Tourism Planning: A Self-assessment Instrument." *Annals of Tourism Research* 31, no. 3: 623–39.

Richards, Greg. 2011. "Creativity and Tourism: The State of the Art." *Annals of Tourism Research* 38, no. 4: 1225–53.

Richards, Greg, and Derek Hall, eds. 2003. *Tourism and Sustainable Community Development*, vol. 7. London: Routledge.

Richards, Greg, and Julie Wilson. 2006. "Developing Creativity in Tourist Experiences: A Solution to the Serial Reproduction of Culture?" *Tourism Management* 27, no. 6: 1209–23.

Roberts, Les. 2012. "Cinematic Cartography: Projecting Place through Film." In *Mapping Cultures*, edited by Les Roberts, 68–84. London: Palgrave Macmillan.

SCIWAY. n.d. "Movies Filmed in South Carolina—Dear John." Accessed 6 April 2018. https://www.sciway.net/movies/sc-movie-dear-john.html.

Stake, Robert. 2000. "Case Studies." In *Handbook of Qualitative Research*, edited by N. K. Denzin and Y. S. Lincoln, 435–53. Thousand Oaks, CA: Sage.

Tanzi, Christine. 2017. "Pablo Escobar Slept Here: Too Soon for Narco-Tourism?" *Bloomberg Businessweek*, 14 November. https://www.bloomberg.com/news/features/2017-11-14/pablo-escobar-slept-here-is-it-too-soon-for-narco-tourism.

Timothy, Dallen J. 2011. *Cultural Heritage and Tourism: An Introduction.* Bristol, UK: Channel View Publications.

Timothy, Dallen J., and Cevat Tosun. 2003. "Arguments for Community Participation in the Tourism Development Process." *Journal of Tourism Studies* 14, no. 2: 2.

Tkalec, Marina, Ivan Zilic, and Vedran Recher. 2017. "The Effect of Film Industry on Tourism: Game of Thrones and Dubrovnik." *International Journal of Tourism Research* 19: 705–14.

Visit Albuquerque. n.d. "Film Tourism." Accessed 6 April 2018. https://www.visitalbuquerque.org/about-abq/film-tourism/.

Volo, Serena, and Anna Irimias. 2016. "Film Tourism and Post-Release Marketing Initiatives: A Longitudinal Case Study." *Journal of Travel & Tourism Marketing* 33, no. 8: 1071–87.

Ward, Susan, and Tom O'Regan. 2009. "The Film Producer as the Long-Stay Business Tourist: Rethinking Film and Tourism from a Gold Coast Perspective." *Tourism Geographies* 11, no. 2: 214–32.

Wray, Meredith, and W. Glen Croy. 2015. "Film Tourism: Integrated Strategic Tourism and Regional Economic Development Planning." *Tourism Analysis* 20: 313–26.

Art Worlds in the Periphery: Creativity and Networking in Rural Scandinavia

Solène Prince, Evangelia Petridou, Dimitri Ioannides

Introduction

Creativity has become a popular strategy for promoting the innovations and ideas behind economic growth (Richards 2011; Ray 1998). The urban-centric rhetoric behind creativity is, however, problematic when theoretically and practically applied to peripheral spaces despite instances of emerging creativity in the countryside and small cities in peripheral regions (Cloke 2006; Gibson 2010). Overall, rural cultural and creative clusters face many developmental challenges because they are peripheral with respect to global markets, happenings, and publics (Andersen 2010; Gibson, Luckman, and Willoughby-Smith 2010). Thus, various scholars have called for policies and theories that take into account the nature of creativity and networking in rural areas, while concurrently avoiding urban-centric rhetoric as a means of studying, planning, and assessing the success of rural clusters (Cole 2008; Fløysand and Jakobsen 2007).

In this chapter, we offer insights into the development of such arrangements and theories that demonstrate enhanced sensitivity to the nature of creativity and networking in rural areas, especially in light of these regions' peripherality, as a result of which reliance on public funding and tourism development are commonly used to counter limited economic and social opportunities. Unlike in cities, the development of creative industries in rural areas often stems from cultural and artistic clustering embedded in traditional products, narratives, practices, and livelihoods (Bell and Jayne 2010). With the decline of primary industries, many rural areas have reinvented themselves as sites of appealing localized consumption (Everett 2012; Mitchell 2013). There is, thus, often a heavy reliance on tourism to maintain emerging rural creative industries and the appealing atmosphere they foster (Cloke 2007). Moreover, there is consensus that the development of creative clusters in the periphery should aim at creating better living spaces by improving territorial cohesion (Petridou and Ioannides 2012).

Our interest in this chapter is the various types of networks that emerge in the rural space as artists pursue their own professional ideas, but also promote community development, which has the potential to counter the effects of limited economic and social opportunities in the periphery. Our study object within the broad field of creativity, as defined by Richards (2011), is the creative space. He conceptualizes this space as clusters of creativity or culture, which are "perhaps the most obvious physical manifestation of the relationship between tourism and creativity" (1240). Our cases reflect the importance of acknowledging that, while not all actors involved in rural development are tourist entrepreneurs per se, they nonetheless play a significant role in positioning the periphery as an interesting place to live in and visit. In the first case, we present a permanent network. We identify this network as a significant factor behind the reconciliation of tourism and professional development in the periphery. In the second case, we investigate an ephemeral network, which we identify as fundamental to strengthening the concept of territorial cohesion among various residents, and to turning the periphery into a site of creativity of potential interest for visitors and tourists.

The cases we present are from Scandinavia and highlight two of the themes Scherf identified as central to this volume in the introduction. Our first case, the craft artists of the Arts and Crafts Association Bornholm (ACAB) in Denmark, relates to the theme of creative networks offering touristic experiences. Through EU funding, these artists have formed an association to take advantage of the tourist season by exchanging skills and pooling resources, but also by creating a local identity based in idyllic rurality. Additionally, these artists go abroad individually but also organize biannual meetings in Bornholm to foster inspiration and learning. We also examine the project of Konstgödning (Art Fertilization) in Jämtland, Sweden, which relates to the (re)generation of sustainable cultural development in host communities. The diverse artists involved in this project received money from the EU in 2015 to co-produce art that would highlight locals' relationship to their peripheral location in subarctic Sweden. The project outlines the development of creative output based in territorial cohesion stemming from ephemeral regional networking. It created space to foster a series of events based in the creativity of local artists and the identity of residents. Both cases underline the highly embedded nature of rural creative networks, leading us to encourage the careful study of these networks to promote sustainable tourism development in the periphery.

Creativity and Spatial Development

Cities and regions around the world increasingly emphasize creativity as a popular strategy for encouraging innovations and ideas for fostering economic growth (Richards 2011; Ray 1998). The application of creativity to planning and development strategies results from broader global processes leading to widespread commodification and standardization, the rapid development of the knowledge and network economies, and rising competition between cities and regions to attract both visitors and residents (Pratt 2008; Scott 2010a). Although creativity is hard to define (Richards 2011), it is generally perceived to imply inventiveness, displays of imagination, and stepping beyond traditional ways of making, knowing, and doing (Richard and Wilson 2007). These elements manifest themselves differently within creative spaces through the presence of creative persons, such as artists and designers; the development of creative production processes; the appearance of creative products, such as consumer experiences; and the flourishing of creative environments (Florida 2005, 2002; Richards 2011; Scott 2010a).

The appearance of creative production processes relates to the development of creative industries, which are seen as "activities which have their origin in individual creativity, skills and talent, and which have a potential for wealth and job creation through the generation and exploitation of intellectual property" (Richards and Wilson 2007, 5). The pioneering research of Richard Florida (2005, 2002) and Charles Landry (2000) has inspired policies aimed at reinforcing the appearance of urban neo-bohemias, where deindustrialized facilities house small firms working in, for instance, design, film, music, or publishing. This research stresses the influence of members of the so-called "creative class," who Florida (2005, 2002) believes are primarily drawn to particular spots in an urban environment precisely because they offers an ideal combination of attributes, which inspire further creativity (see also Currid 2007; Zukin 2010).

Market competitiveness within the creative economy stems from the formation of those attractive neighbourhoods that inspire the development of clusters of creative entrepreneurship (Scott 1999; Watson 2008). Subsequently, creative clusters promote the development of the kind of atmosphere that entices not only tourists but also new residents. As Mommaas (2004) explains, creative clusters strengthen their host communities' identity, attraction power, and market position. They are, in this sense, hubs of prosperity, cosmopolitanism, novelty, and leisure. Within these trendy environments, specific products of the creative industries provide tourists with experiences through the development of events, cultural routes, and themed spaces (Richard and Wilson 2007). In recent years, although increasing academic criticism has been levied against the work of Florida and Landry (see, for instance, Pratt 2008; Scott 2010a; Sorensen 2009), the importance of creativity and its related cultural industries for urban and regional growth remains largely uncontested at the policy level. As Scott (2010a) contends, what planner or politician would not want their city or town to be vibrant, attracting visitors and residents alike?

Research on creativity has traditionally been heavily biased toward cities (Gibson 2010). Nonetheless, researchers increasingly describe various creative industries and initiatives flourishing in smaller cities and rural areas (see, for instance, Fitjar and Jøsendal 2016; Gibson 2010; Petridou and Ioannides 2012; McGranahan and Wojan 2007; Waitt and Gibson 2009). The development of rural creative industries generally stems from cultural clustering embedded in spatial symbols, practices, and attributes, such as traditional products and livelihoods (Bell and Jayne 2010). Scott (2010b) contends that the creative development of rural areas often derives from the landscape's physical and cultural features. He presents the English Lake District as an example of such diverse place-based practices, ranging from gastronomy and handicrafts to outdoor activities. Diversity also features in the study from Harvey, Hawkins, and Thomas (2012) in

Cornwall, England, where a broad range of creative individuals such as designers, visual artists, and performing artists have clustered. When actual subsectors of creative clusters emerge in rural areas, they often stem from the development of arts-and-crafts hubs in amenity-rich areas (Wojan et al. 2007).

To sustain creative industries in rural milieus, policy-makers frequently rely heavily on tourism (Cloke 2007). With the decline of primary industries, many rural areas have reinvented themselves into sites of consumption (Everett 2012; Mitchell 2013; Sims 2009; Zasada 2011). Governmental attempts at fostering development in rural spaces have long targeted the promotion of place-based products to preserve cultural integrity and deliver unique competitive products (Kneafsey 2001; Ray 1998). Objects and symbols of peripherality and rurality infuse rural spaces and products with their own authenticity (Connell and Gibson 2003; Stratton 2008). Subsequently, tourists in rural places consume these mostly to satisfy a need for authentic experiences; in such environments, close encounters with local people and new sensations matter (Sims 2009). This search for authenticity is especially relevant for city dwellers who feel increasingly alienated from the production processes behind their consumption patterns (Mitchell 2013).

In light of these challenges and opportunities, we explore in the rest of our theoretical section two key components behind the creative development that makes the periphery attractive for both tourists and artists. First, we take up the spatial character of networking as it as a fundamental aspect of creativity and occurs significantly differently in the periphery compared to the urban centre. Secondly, we look into the social significance of networking for creative clusters in the periphery by bringing up territorial cohesion.

Networks and Creative Development

Bell and Jayne (2010) argue that the urban-centric rhetoric of creativity means that policies concerning creative industries and initiatives are generally not suited to the realities of rural areas. Creativity is easy to find in cities, leading many researchers to assume that developmental paths focusing on the creative industries can flourish elsewhere by following a similar plan (Gibson 2010). Fløysand and Jakobsen (2007) claim that this urban-centric rhetoric usually tends to favour chic trends better suited for cities, thereby undermining the more traditional nature of the creative practices of rural areas. Mostly, familiarity and social roles foster different environments of creative development (Atterton 2007; Scott 2010a). Informal networks, often highly embedded in the local context, are commonly found in peripheral locations. Harvey, Hawkins, and Thomas (2012) hold that the informal environment represents a work-based support system serving, for instance, to oversee emotional needs and provide professional advice while enabling the pooling and sharing of resources.

Most importantly, perhaps, networks in rural areas are characterized by high density; that is, tight structures with actors connected through strong ties (Tregear and Cooper, 2016). Dense networks are cohesive and can be visualized, according to Borgatti, Everett, and Johnson (2013), as a big clump of tangled electrical wires. Network cohesiveness corresponds closely to social cohesiveness, though this is not necessarily always the case. Density is a term best used comparatively, and the density of small networks, which one might expect to find in rural spaces, tends to be higher than the density of larger networks, which may, more likely though not exclusively, be found in urban areas (Borgatti, Everett, and Johnson 2013). A disadvantage of dense networks is that they do not benefit through the advantages of weak ties. Granovetter (1973) notes that "weak ties . . . are here seen as indispensable to individuals' opportunities and to their integration into communities; strong ties, breeding local cohesion, lead to overall fragmentation" (1378). In places characterized by limited social mobility and lifelong friendships, actors tend to stay locked in a certain mindset and harbour feuds because they have always done so (Atterton 2007).

The nature of creativity implies the formation of networks to inform and inspire its key instigators. It also requires the presence of an audience to make these creations meaningful (Scott 2010a). This is why clusters in urban areas are often seen as the epitome of the creative space. However, density is not a defining element behind the success of a creative cluster, as it can be overcome through temporary co-location (Cole 2008; Comunian 2017; Harvey, Hawkins, and Thomas 2012; Norcliffe and Rendace 2003; White 2010). Harvey, Hawkins, and Thomas (2012) show that rural artists and creative practitioners build ephemeral networks with other actors through programs such as residencies, workshops, and conferences, keeping them in touch with extra-local colleagues' work and the latest developments in their field. These practitioners pursue extra-local connectivity with relevant creative practitioners, but also intermediaries, producers, and consumers. Comunian (2017) explores temporary networking at a street-art festival in Medway, UK, where artists attend the event to learn practical and artistic skills, expand their social and professional networks, and find a forum to create innovative work. Similarly, Duxbury, in chapter 1 of this volume, shows that networking is essential among creative tourism entities in smaller places for purposes of visibility, knowledge sharing, and capacity development.

Territorial Cohesion and Creative Development

Rural creative clustering often occurs in amenity-rich areas, such as mountain or seaside resorts. These places are especially attractive to creative individuals seeking an ideal rural lifestyle of tranquility, inspiration, and, perhaps, outdoor activity (McGranahan and Wojan 2007; Wojan et al. 2007). For the individuals who form these clusters, economic success is often not the main driver affecting their choices of location and lifestyle (Comunian 2009; Petridou and Ioannides 2012; Prince 2017). Rather, there is a desire to combine the artistic lifestyle with participation in outdoor recreational activities, flexible working hours, off-season tranquility, and inspiration from the physical environment (Prince 2017). Artists often foster strong objectives of artistic and personal integrity that can clash with the needs of profit-driven business (Comunian 2009; Deener 2009; Prince 2017; Sheehan 2014). It is not unusual for artistic development and lifestyle values to take precedence over strategic business development, like marketing and finance. In this regard, informal social networking and personalized encounters with customers and the public become key elements behind the identity formation of peripheral creative clusters.

The urban rhetoric of creative development overlooks that rural areas face specific challenges when developing their cultural industries into assets of the creative economy. Researchers have identified many of the developmental challenges creative and cultural clusters face due to their peripherality vis-à-vis global markets. For instance, Bennet (2010) and Davies (2008) highlight that urban centres mostly attract and retain creative agents to the detriment of peripheral areas, robbing the latter of its creative capital. The social perception and reality of peripheral rural areas is thus often one of desolation and of limited career opportunities. In chapter 6, Aquino and Burns identify this issue in the Icelandic context, adding that this social perception stems from a power imbalance that diminishes the legitimacy of local knowledge in the face of the scientific knowledge usually associated with urban areas. Such perceptions and local conditions emerge as serious obstacles to encouraging in-migration and economic development, which are crucial to creative clustering. The high cost of reaching urban centres for business exchange, professional communication, and social networking strains the socio-economic development of peripheral clusters and industries (Andersen 2010; Gibson, Luckman, and Willoughby-Smith 2010).

Significant to the socio-economic development of the periphery, the social objectives of artistic integrity often transcend to the integrity artists demonstrate toward place and society. For instance, Petridou and Ioannides (2012) explain that artists can promote the development of community meeting spaces through their artistic practices, as a means of strengthening social bonds and feelings of belonging in peripheral areas. Such a practice is an example of territorial cohesion, one of the European Union's three cohesion policies (the others being economic and social cohesion). Essentially, territorial cohesion promotes the concept that no one should be disadvantaged because of where one lives in the union. Small-scale, artistic arrangements, like those examined in this chapter in the Swedish periphery, help disadvantaged places realize their developmental potential through an inward-looking exercise fostering local heritage preservation, social entrepreneurship, collective action, and inclusivity. All these are constituent elements of territorial efficiency, territorial quality, and territorial identity, which are in turn dimensions of territorial cohesion (Camagni 2007).

Case Studies

Rural creative clusters emerge and survive through various forms of institutional support, public and private partnerships, and training organizations (Harvey, Hawkins, and Thomas 2012). In Scandinavia in general and in Sweden in particular, national cultural policy has been tied to sustainable regional development, whereby "an attractive region with developed natural, cultural, and cultural heritage resources contributes added value to the business environment and furthers the regional competitive power" (Lundström quoted in Petridou and Ioannides 2012, 128). While these agendas usually prioritize the growth of creative industries for regional economic development, the aim of territorial cohesion also encompasses development schemes. Financing schemes differ in nature and scale, though in Scandinavia currently such financing is packaged as limited-time project funding. The two cases we explore represent networks of artists in Scandinavia born out of such policies, objectives, and projects.

For both cases, we employed qualitative fieldwork to collect data. A qualitative approach is useful when seeking to understand different facets of social life (Creswell 2013). It is important for the qualitative researcher to look into stories, encounters, daily practices, and visual and material assemblages at the heart of local experiences. In the case of Bornholm, the main author conducted two sets of fieldwork consisting of interviews with nineteen of the sixty-four members composing the ACAB at the time of the study in the fall of 2014 and 2015. She also employed participant observation at different venues and events relating to arts and crafts. This diverse sample includes ceramists, textile and glass designers, a potter, and a woodworker, each from various locations around Bornholm. In the case study relating to Jämtland, the two co-authors followed the project Konstgödning (Art Fertilization) from inception to completion. They kicked off the data collection with a workshop with the two project managers, who were also artists, in the fall of 2015. They collected further qualitative data through observations, in-depth interviews, and email correspondence with the nine artists involved in the project over the period of one year. They attended the meetings between the project managers and the artists, as well as the public closing event in September 2016.

The Danish network emerged in the context of a more general branding of Bornholm as a creative place, whereas the Swedish network was ephemeral, the result of project financing in a region that prides itself on its winter sports profile. Despite the differences between the networks themselves, the cases share similarities. Both are situated in the Scandinavian periphery, in sparsely populated regions that had to reinvent themselves in the 1990s due to widespread economic restructuring. In the case of Bornholm, this had much to do with the gradual disappearance of a once flourishing fishing sector, while in Jämtland it was due to the closing of military bases and the downsizing of various traditional industries.

The Craft Artists of Bornholm: Reconciling Tourism and Professional Development

The case of the ACAB in Denmark is an example of the type of networking fostered through an agenda of creative and cultural development for economic purposes. Bornholm is a 588-square kilometre island in the Baltic Sea with a population of 39,584 (Danmarks Statistik 2020). It underwent major economic restructuring in which, starting in the 1990s, several European and national programs were implemented in an attempt to counter the collapsed of the island's economy. Many of these incentives aimed at boosting the tourism industry. The restructuring stimulated the development of various micro-businesses involved mostly with specialized foods, hospitality, and handcrafts, which now characterize Bornholm's economy and destination brand (Ioannides and Petersen 2003; Manniche and Larsen 2013).

With annual visitor numbers at around 750,000, tourism is a major source of earning for the island. The vast majority of these tourists arrive within a short summer season beginning in June and ending in September. The rural idyll is significant in the construction of Bornholm as a summer destination. With its quaint fishing villages, sandy beaches, and family-friendly cycle paths, the island represents for many of its visitors simple living in the countryside (Larsen 2006). The island's cultural appeal also stems from its arts-and-craft cluster. The ceramics tradition on Bornholm has its origins in the clay pits of the Middle Ages and clay factories of the nineteenth century, but it is in the 1930s that potters started opening private studios, redefining their aspirations to work with clay as artists. The glass-making tradition stems from the entrepreneurship of a Danish businessman who in the 1970s used old smokehouses to fit furnaces to work with glassware. Since then, other types of art studios have flourished on Bornholm, where craft artists work with textile and wood, among other materials.

In the early 2000s, the arts-and-craft cluster garnered attention from local politicians who were seeking to boost tourism on Bornholm through EU funds for regional development. The strategy was to brand each of the island's regions and, thus, a decision was made that the Hasle municipality would focus on arts and crafts. This sparked the grassroots formation in 2002 of the ACAB by a small group of local craft artists. Their goal was to come together in order to facilitate and increase their individual and collective ability to profit from the tourist season. ACAB membership rests on a strict selection process to preserve high-quality standards within the group. Only professional artists residing permanently on Bornholm can become part of this network. The ACAB eventually became a significant body representing the interests of the craft artists in local politics. For instance, it has a member representing their interest in the Local Action Group dedicated to distributing EU funds on Bornholm. The ACAB exists to help its members increase their earnings by committing to preserve artistic quality among the group, promote its visibility around and beyond Bornholm, and help it foster the artistic, administrative, and commercial skills its members need to succeed. As an association, the craft artists are able to apply for all sorts of funding from national and transnational bodies.

Forming a Permanent Network

The ACAB members see their association as a platform for internal co-operation with which they can support each other in different ways without feeling threatened by competition. One of the founding members acknowledged that their association encouraged co-operation within the cluster in a way that did not exist before:

> We had the organization and we were seeing each other. . . . I did not know them all, but suddenly little groups started working together. In that way, suddenly you were getting so much more out of everything because we worked together in an association.

The cohesion of the group also comes from its ability to pool resources that individuals often find difficult to access. It was thus crucial for these craft artists to find ways to maximize their creative time. For instance, the ACAB exhibits its members' crafts every summer during the tourist season at a place called Grønbechsgård in Hasle. These artists recognize that this touristic venue brings them customers who develop an interest in their art after seeing their work on display there. Moreover, the ACAB membership fees and other funds pay for a secretary who is dedicated to servicing the aims and ambitions of the members by finding and applying for funds, booking exhibitions, and more. By assuming these administrative tasks, the secretary can allow the artists more time to indulge in their creative endeavours.

Despite the fact that many craft artists seek profits during the tourist season, they remain skeptical as to whether the destination-management organization for the island—Destination Bornholm—responds to their individual needs as professional artists. As many of them pointed out, it is expensive to be a member of Destination Bornholm, an organization that fosters strategies geared more at promoting the natural landscape and family holidays. The ACAB is, of course, more apt at servicing the interests of the craft artists because it recognizes that their strategies for taking advantage of the tourist season unfold in the wake of their professional and artistic ambitions. The ACAB is a mechanism aimed at preserving quality in the arts-and-craft brand on Bornholm in light of amateur competition. Many craft artists explained that they find bargaining power in the discourse of quality and locality the ACAB imposes with its entrance criteria. One ceramist mentioned the distinction the ACAB membership brings to her art:

> It's what you want. It's not a hobby, it's serious. It's quality. It's professional. It's a stamp for this is the good art and there is a difference. But if the ACAB wasn't there they [the tourists] wouldn't know.
>
> I have many people who actually say: "We went to many places and it was all [bad]. We were so tired we didn't want to go in any more places because we were getting so disappointed." Then, I say, "I have to tell you that we have the ACAB. We have different levels [of competence] here for people who make these kinds of things."

These entrance criteria for the ACAB work to define the crafts of its members as special compared to the work of amateurs or of serial reproduction.

The ACAB diffuses a brand of creativity that reinforces internal cohesion since the craft artists commit to uphold Bornholm's identity as a special place in their crafts due to their professionalism. The networking reinforces the appeal of the creative space, where many craft artists were found to believe that these creative and spatial characteristics are precisely what attracts tourists to their crafts.

Temporary Networks Beyond the Island

Working collectively has also facilitated networking beyond the insular cluster, enabling the members of the ACAB to remain connected to the wider world of arts and crafts. It was clear that those extra-local networks primarily fulfill a need to find inspiration for individual artistic purposes. Many of the members of the ACAB travel a lot as individuals, supported by different sources of funding, attending residencies and exhibitions worldwide. The island hosts biennial glass and ceramics conferences to which are invited craft artists from Europe who share their skills and inspiration through exhibits, workshops, and lectures. Also, Bornholm is home to the Danish School of Design, which teaches glass and ceramic art and receives artists in residence.

Importantly, extra-local networking reinforces the internal cohesion of the ACAB as its members present themselves as a group with collective aims who face particular challenges in relation to their peripheral location. It became apparent that creative individuals from the periphery connect with other creative individuals from the periphery to share experiences as much as creativity. For instance, the ACAB has specifically worked to form networks with art groups from Nordic islands facing similar challenges to those seen in other peripheries, like Gotland and Öland in Sweden, and the Faroe Islands. ACAB members have also visited Greenland and Iceland to widen their networks in northern Europe. One of the founding members explained that, as the co-chairperson, she once visited with her fellow chairperson a place in the far north of Iceland to speak about the success of the ACAB, as artists there wanted to form their own association based on Bornholm's success. Thus, creative individuals in the periphery also network to support each other and address strategies related to the social and spatial challenges they face.

Konstgödning in Jämtland: Fostering Territorial Cohesion

The second case, the project of Konstgödning in Jämtland, Sweden, outlines the development of creativity and networking through the agenda of territorial cohesion. The artists involved in this project received money from a national organization in 2015 to carry out an art project highlighting the local people's relation to their peripheral location in subarctic Sweden. Konstgödning was the umbrella project for nine subprojects in the county—one for each municipality, with one having two subprojects. The project demonstrates the development of creative output based in territorial identity stemming from regional networking. It aimed to develop the linkages between art and place and, specifically, the in situ co-production of art by the inhabitants of each municipality with selected artists. Two project managers received the funding and then advertised a call for nine artists, each willing to work with one municipality. All but one had a connection with the municipality they chose; either they came from their chosen municipality or had lived there in the past. In this way, project offered the artists a way to reconnect with Jämtland.

Jämtland is a large, sparsely populated county covering about 12 per cent of the national territory, though it is home to just 1.3 per cent of the Swedish population (amounting to approximately 130,810 people) (Region Jämtland Härjedalen 2019). Östersund is, by far, the largest municipality, with approximately 58,000 residents. Traditionally, the region has been heavily dependent on primary extractive activities, while the area in and around Östersund has had a heavy military presence. The shutdown of the military bases in the early 1990s led to a series of policies aimed at encouraging economic diversification, including the establishment of one of the campuses of the country's newest university (Mid-Sweden University) and the opening of several central government facilities in and around Östersund. Meanwhile, tourism has been promoted as a major activity, generating about SEK2.5 billion per year and employing 3,000 year-round workers. The number of bed-nights in the county is just 9 million per year. The principal destination is Åre, one of Sweden's leading winter sport resorts, which is increasingly diversifying into a year-round destination. Meanwhile, other destinations, including several rural settlements, struggle to attract visitors. Even though Konstgödning was never meant to entice visitors, creative projects such as this can serve to give new life to some rural places, making them destinations in their own right.

The objective of the co-production of art in situ was that rural inhabitants would be able to feel that positive energy was being produced; that their rural realities were seen as important and deserving attention; and, finally, that the act of producing art would make a positive difference to these realities. In other words, the main purpose of the project was to engender dialogue about place, about what it means to occupy space in the outskirts of Europe, and to bring increased attention to the realities of uneven development rather than explicitly be a means for economic development or produce art for art's sake. The production of art is used as a vehicle to highlight the positive elements of each place instead of pinpointing the usual complaints about living in a peripheral place, such as bad connectivity to the capital and lack of services of general interest. Given the space restrictions and the scope of this chapter, three of the nine subprojects are presented below as examples of the forms of networks created for territorial cohesion that turn the periphery into an attractive creative space.

Networks to Connect Residents

The subproject of the Municipality of Bräcke, in Sundsjö, sought to help residents come together to discuss issues important to them as inhabitants of a peripheral area. The project was managed by Vanja Steinholtz and involved local residents forming a choir as well as writing and performing music during a summer festival event in 2015. Steinholtz, a young musician in her twenties, was the only artist without any direct personal connection to the municipality in which she worked. Therefore, she had to contend with locals' perceptions of her as young Stockholmite expert-artist imposing her know-how on the non-expert country folk—a core-periphery power imbalance that is common in the area. She was aware that she had to maintain a balance between guiding and imposing while allowing people to express themselves without confining them to the "one right way" of writing and performing a music piece.

The choir was formed through *conversation workshops*, in which residents formed groups and talked about things important to them. The topics of discussion revolved around place: what people would miss if they moved away, what they miss where they live now. The same issues kept coming up; sometimes the discussion became quite political, with participants starting to talk about how rural areas are an afterthought in Swedish politics, how meaningful and traditional jobs disappear only to be replaced by jobs at facilities that serve primarily national interests and impact the local landscape negatively, such as nuclear power plants and wind turbines. Even so, there was more to the conversation than expressing dissatisfaction about rural economic decline and exploitation. For instance, in one workshops in late May 2015, Steinholtz described how participants talked about the beauty of the surroundings—"connecting to place in a lyrical and emotional fashion." These meetings thus created a space for residents to connect positively over the attributes of their landscape and what they cherished about it in light of, rather than despite, its peripherality.

Though the choir performance and its creativity was the focal point of the festival, which took place in Tavnäs on 15 August 2015, a host of local artisans also participated, including artists and local food producers. The event was featured as a summer happening and it attracted not only locals, but also tourists from the surrounding area. Having said this, even though Steinholtz had hoped that the temporary network of inhabitants writing music and singing together would continue meeting, this has not materialized.

Networks to Connect with Asylum-Seekers

The subproject of the Municipality of Härjedalen, in Hede, sought to create meeting spaces where newly arrived asylum-seekers could gather with established Hede inhabitants. Whereas the asylum-seekers came from Eritrea and Syria, the long-term residents were ethnic Swedes. In 2015 and 2016, Sweden received an influx of asylum-seekers in the wake of the terror engendered by the Syrian Civil War and political unrest in East Africa. The Swedish government has a policy of locating a number of asylum-seekers in sparsely populated areas of the country, which raises issues related to the asylum-seekers' integration into Swedish society. The project manager, Bodil Halvarson, turned to a group of local Swedish language instructors for help in recruiting individuals from the immigrant community. She contacted the church and the local village association for help in recruiting residents. Because Halvarson's family comes from the area, she also enlisted friends and relatives in this effort. The result was fourteen participants in a series of meetings between recent immigrants and long-time Hede residents, each lasting approximately two or three hours. As Halvarson reported, "I wanted to explore how art as a tool might interest people to share their stories with each other and [how] by taking an active, physical, and creative part in each other's stories, [they] could become part of each other's life and memory."

Each of the participants produced a piece of art with the larger aim being to get people talking. Halvarson provided a set of questions as general guidance; some participants followed while others did not. Notably, she did not actively participate in the meetings, but instead sat in the back and took notes as none of the participants transcribed the meetings. It was the voice of the recent arrivals and the long-time residents that she wanted people to hear; as such, she functioned as a facilitator, an amplifier.

One would perhaps expect that the resulting narratives would mostly come from the asylum-seekers, and yet the co-production of art becomes the grammar of mutual storytelling. Hede residents, both relatively recent arrivals and those born and raised in the town, came together to share their stories and experiences. One long-time resident spoke of how he worked in the industry when he was young, how he hurt his back and had to find the strength to retrain for a second career. These stories, along with the art produced during the meetings and the photographs of the participants, comprise an exhibition that is being displayed in local libraries. This exhibition has thus become an element of the attractive creative rural space, offering residents and visitors alike something interesting to experience in the periphery. Halvarson has presented her project to different audiences outside events organized by Konstgödning. What is more, she forged friendships with the project participants. Halvarson herself is haunted by the tiny green clay boat Hussein made to tell his story. She has exhibited the artifact in prominent places, along with the rest of the art that was produced during the meetings, on several occasions.

Networks to Connect Women

Martin Johansson drew from his experience as an artist and a director for his project on women's stories in Kluk in the Municipality of Krokom.

The project consisted of a series of meetings in Kluk, where local women told their stories to each other. Johansson wanted to amplify Kluk's female voices and to distinguish them from the masculine hunting-and-fishing image with which the area is largely associated. It is important here to note that the storytelling was intended mainly for the consumption of the participants. There was a sold-out event in Kluk on 23 May 2015, but otherwise these women would not come to Östersund, the county capital, to perform. If one wanted to listen to them, then one had to also visit the space they inhabited. This creativity came from the rural periphery and was only to exist there. This pride of place, evident in the women's narratives, lent a certain urban ephemerality to the event since no official video exists—like, say, a Springsteen concert in Gothenburg or a Madonna event in Melbourne. Rather, one had to be there. What does exist, however, is a book with the women's stories.

At these project meetings, the women of Kluk sat around the table and talked; some spoke in the local dialect. They discussed their families, where they come from, trips they took. One woman speaks of how her mother died in the late 1940s, when she was four weeks old, and how her father had to raise her and the half-brother from her mother's previous relationship while also taking care of the farm all by himself. Another cites her journal entry from a trip to Kenya in which, in a way that is all too familiar to those who journal, she apologizes for not having written recently—it had been too hot. A third says she grew up in Springfield, Missouri, which incidentally is the city in which two of the authors of this this chapter lived for a large portion their adult lives. These narrative snapshots provided a sense of the distinct voices and connections that can come out of the act of talking about oneself, which is not exactly self-evident in the Swedish periphery, where the closest neighbour can be miles away. The women forged bonds that have outlived the May 2015 event, in line with the goals of the project in general and Martin Johansson's constituent project in particular.

Conclusion

In the introduction to this volume, Scherf suggests that, through creative tourism, all places have the ability to use their unique characteristics to attract visitors. We have presented different types of rural networks that emerge through the collaborative work of artists. These artists are able to draw on cultural, creative, and artistic capital to offer experiences, products, and events to visitors and residents alike. While the cases examined in this chapter highlight different purposes and dynamics of artist networks, both feature networks that are highly embedded in places where creative individuals cluster. On the one hand, rural artists can create networks to profit from their art, develop competences, and find inspiration in light of their remoteness from buyers and happenings. The ACAB, as a permanent network, has enabled the craft artists of Bornholm to not only overcome their distance from urban centres, but to preserve an identity based in what they value about their rural island as artists: the handmade mode of production, close encounters with buyers, and the local character of their art. Temporary networks are important for finding other craft artists from the periphery with whom to share best practices and inspiration. These networks enable the craft artists of Bornholm to take advantage of the tourist season as they pool resources to strengthen and diffuse a viable brand based in culture and creativity that appeals to visitors.

On the other hand, we have shown artists to be significant actors when it comes to enhancing territorial cohesion in the wake of the spatial disadvantage that imprints negative perceptions on space. The Konstgödning's subprojects functioned as the impetus for ephemeral networks, bringing people together to tell stories about their lives and the space they inhabited. This socially inclusive activity aimed at looking inward and bringing forth positive aspects about life in the periphery, giving a voice to anyone wanting to speak. Concurrently, the impact of projects like this are untraceable and cannot easily be quantified. Art, or rather the creation of art, opened up a space for residents in general, but also more specific groups of residents such as women, asylum-seekers, and artists of all sorts, who joined in this project in order to share images of rurality and snapshots of their lives within it. The output of the subprojects was formally a series of creative events, potentially attracting tourists and visitors from the outside to the extent that it was part of the summer program of cultural happenings in the county. Despite its ephemerality as a series of one-and-done events, the project fulfilled a more fundamental need among the county's inhabitants for cultural sustainability.

Artist networks have the potential to turn peripheral spaces into creative domains, making them attractive places in which to live, and thus to visit. These networks are themselves the result of collaboration between the public and private sectors, and institutional support and various partnerships and training organizations are crucial to the sustained existence of rural creative clustering (see also Harvey, Hawkins, and Thomas 2012). We contend that the rural creative space that attracts visitors will be the one where there is support for the development of creative networks that enables its members to take advantage of their own embeddedness in the rural milieu. Scholars recognize that rural creative industries are embedded in spatial symbols, practices, and attributes, such as traditional products and livelihoods (Bell and Jayne 2010; Scott 2010b). By studying rural artist networks, we were able to map out some of the dynamics behind the development of creative spaces in the periphery.

We argue that rural creative tourism development is closely linked to the networks, both permanent and temporary, that facilitate the commercialization of art, foster artistic inspiration and best practices, and pursue social objectives to establish the periphery as a lively space. It is thus important to carefully study these processes before outlining any type of recommendation for the sustainable development of rural areas through creative tourism. Networks consisting of actors from the public and private sectors are increasingly important when it comes to tourism-related policy-making (Dredge 2006). Given the industry's fragmented nature as a societal phenomenon and an economic activity, relational structures that bring people together formally and/or informally are even more salient. Networks can be conceptualized as structures that embed local values and meanings and that "over time become regimes of power and knowledge that operate to filter, prioritize and promote particular local tourism policy actions and initiatives" (Beaumont and Dredge 2010, 8). These tourism policies and initiatives will have to consider the issue of scale so that small and sparsely populated areas are not pushed into fostering large-scale events and projects that do not consider the embeddedness of the products and stories diffused to tourists interested in the rural periphery.

References

Andersen, Lisa. 2010. "Magic Light, Silver City: The Business of Culture in Broken Hill." *Australian Geographer* 41, no.1: 71–85.

Atterton, Jane. 2007. "The 'Strength of Weak Ties': Social Networking by Business Owners in the Highlands and Islands of Scotland." *Sociologia Ruralis* 47, no. 3: 228–45.

Beaumont, Narelle, and Dianne Dredge. 2010. "Local Tourism Governance: A Comparison of Three Network Approaches." *Journal of Sustainable Tourism* 18, no. 1: 7–28.

Bell, David, and Mark Jayne. 2010. "The Creative Countryside: Policy and Practice in the UK Rural Cultural Economy." *Journal of Rural Studies* 26, no. 3: 209–18.

Bennett, Dawn. 2010. "Creative Migration: A Western Australian Case Study of Creative Artists." *Australian Geographer* 41, no. 1: 117–28.

Borgatti, Stephen P., Martin G. Everett, and Jeffrey C. Johnson. 2013. *Analyzing Social Networks.* Thousand Oaks, CA: Sage.

Camagni, Roberto. 2007. "Territorial Development Policies in the European Model of Society." In *Territorial Cohesion and the European Model of Society*, edited by Andreas Faludi, 129–44. Cambridge, MA: Lincoln Institute of Land Policy.

Cloke, Paul. 2007. "Creativity and Tourism in Rural Environments." In *Tourism, Creativity and Development*, edited by Gregory Richards and Julie Wilson, 37–47. New York: Routledge.

Cole, Alexander. 2008. "Distant Neighbours: The New Geography of Animated Film Production in Europe." *Regional Studies* 42, no. 6: 891–904.

Comunian, Roberta. 2009. "Questioning Creative Work as Driver of Economic Development: The Case of Newcastle-Gateshead." *Creative Industries Journal* 2, no. 1: 57–71.

———. 2017. "Temporary Clusters and Communities of Practice in the Creative Economy: Festivals as Temporary Knowledge Networks." *Space and Culture* 20, no. 3: 329–43.

Connell, John, and Chris Gibson. 2003. *Sound Tracks: Popular Music, Identity and Place.* Routledge: New York.

Creswell, John W. 2013, 3rd. edition. *Qualitative inquiry: Choosing Amon Five Approaches.* Los Angeles, CA: Sage Publications.

Currid, Elizabeth. 2007. *The Warhol Economy. How Fashion, Art and Music Drive New York City.* Princeton, NJ: Princeton University Press.

Davies, Amanda. 2008. "Declining Youth In-migration in Rural Western Australia: The Role of Perceptions of Rural Employment and Lifestyle Opportunities." *Geographical Research* 46, no. 2: 162–71.

Deener, Andrew. 2009. "Forging Distinct Paths towards Authentic Identity: Outsider Art, Public Interaction, and Identity Transition in an Informal Market Context." *Journal of Contemporary Ethnography* 38, no. 2: 169–200.

Dredge, Dianne. 2006. "Policy Networks and the Local Organisation of Tourism." *Tourism Management* 27 no. 2: 269–80.

Everett, Sally. 2012. "Production Places or Consumption Spaces? The Place-Making Agency of Food Tourism in Ireland and Scotland." *Tourism Geographies* 14, no. 4: 535–54.

Fitjar, Rune Dahl, and Kari Jøsendal. 2016. "Hooked Up to the International Artistic Community: External Linkages, Absorptive Capacity and Exporting by Small Creative Firms." *Creative Industries Journal* 9, no. 1: 29–46.

Florida, Richard. 2002. *The Rise of the Creative Class*. New York: Basic Books.

———. 2005. *Cities and the Creative Class*. New York: Routledge.

Fløysand, Arnt, and Stig-Erik Jakobsen. 2007. "Commodification of Rural Places: A Narrative of Social Fields, Rural Development, and Football." *Journal of Rural Studies* 23, no. 2: 206–21.

Gibson, Chris. 2010. "Guest editorial—Creative Geographies: Tales from the 'Margins.'" *Australian Geographer* 41, no. 1: 1–10.

Gibson, Chris, Susan Luckman, and Julie Willoughby-Smith. 2010. "Creativity without Borders? Rethinking Remoteness and Proximity." *Australian Geographer* 41, no. 1: 25–38.

Granovetter, Mark S. 1973. "The Strength of Weak Ties." *American Journal of Sociology* 78, no. 6: 1360–80.

Harvey, David C., Harriet Hawkins, and Nicola J. Thomas. 2012. "Thinking Creative Clusters Beyond the City: People, Places and Networks." *Geoforum* 43, no. 3: 529–39.

Ioannides, Dimitri, and Tage Petersen. 2003. "Tourism 'Non-entrepreneurship' in Peripheral Destinations: A Case Study of Small and Medium Tourism Enterprises on Bornholm, Denmark." *Tourism Geographies* 5, no. 4: 408–35.

Kneafsey, Moya. 2001. "Rural Cultural Economy: Tourism and Social Relations." *Annals of Tourism Research* 28, no. 3: 762–83.

Landry, Charles. 2000. *The Creative City: A Handbook for Urban Innovators*. London: Earthscan.

Region Jämtland Härjedalen. 2019. "Befolkning i Jämtlands län." Accessed 13 October 2020. https://www.regionjh.se/regionalutveckling/regionstatistik/statistikefteramne/befolkning.4.67aeab2b16e16577f4fed.html.

Larsen, Jonas. 2006. "Picturing Bornholm: Producing and Consuming a Tourist Place through Picturing Practices." *Scandinavian Journal of Hospitality and Tourism* 6, no. 2: 75–94.

Manniche, Jesper, and Karin T. Larsen. 2013. "Experience Staging and Symbolic Knowledge: The Case of Bornholm Culinary Products." *European Urban and Regional Studies* 20, no. 4: 401–16.

McGranahan, David, and Timothy Wojan. 2007. "Recasting the Creative Class to Examine Growth Processes in Rural and Urban Counties." *Regional Studies* 41, no. 2: 197–216.

Mitchell, Clare J. 2013. "Creative Destruction or Creative Enhancement? Understanding the Transformation of Rural Spaces." *Journal of Rural Studies* 32: 375–87.

Mommaas, Hans. 2004. "Cultural Clusters and the Post-industrial City: Towards the Remapping of Urban Cultural Policy." *Urban Studies* 41, no. 3: 507–32.

Norcliffe, Glen, and Olivero Rendace. 2003. "New Geographies of Comic Book Production in North America: The New Artisan, Distancing, and the Periodic Social Economy." *Economic Geography* 79, no. 3: 241–63.

Petridou, Evangelia, and Dimitri Ioannides. 2012. "Conducting Creativity in the Periphery of Sweden: A Bottom-Up Path towards Territorial Cohesion." *Creative Industries Journal* 5, nos. 1–2: 119–37.

Pratt, Andy C. 2008. "Creative Cities: The Cultural Industries and the Creative Class." *Geografiska Annaler: Series B, Human Geography* 90, no. 2: 107–17.

Prince, Solène. 2017. "Craft-Art in the Danish Countryside: Reconciling a Lifestyle, Livelihood and Artistic Career through Rural Tourism." *Journal of Tourism and Cultural Change* 15, no. 4: 339–58.

Ray, Christopher. 1998. "Culture, Intellectual Property and Territorial Rural Development." *Sociologia Ruralis* 38, no. 1: 3–20.

Richards, Greg. 2011. "Creativity and Tourism: The State of the Art." *Annals of Tourism Research* 38, no. 4: 1225–53.

Richards, Greg, and Julie Wilson. 2007. "Tourism Development Trajectories: From Cultural to Creativity?" In *Tourism, Creativity and Development*, edited by Gregory Richards and Julie Wilson, 89–106. New York: Routledge.

Scott, Allen J. 1999. "The Cultural Economy: Geography and the Creative Field." *Media, Culture and Society* 21: 807–17.

———. 2010a. "Cultural Economy and the Creative Field of the City." *Geografiska Annaler: Series B, Human Geography* 92, no. 2: 115–30.

———. 2010b. "The Cultural Economy of Landscape and Prospects for Peripheral Development in the Twenty-First Century: The Case of the English Lake District." *European Planning Studies* 18, no. 10: 1567–89.

Sheehan, Rebecca. 2014. "Artists' Positioning in and Creating Authentic Identities through Place and Tourism in New Orleans's Jackson Square." *Journal of Tourism and Cultural Change* 12, no. 1: 50–67.

Sims, Rebecca. 2009. "Food, Place and Authenticity: Local Food and the Sustainable Tourism Experience." *Journal of Sustainable Tourism* 17, no. 3: 321–36.

Sorensen, Tony. 2009. "Creativity in Rural Development: An Australian Response to Florida (or a View from the Fringe)." *International Journal of Foresight and Innovation Policy* 5, nos. 1–3: 24–43.

Statistik Denmark. 2020. "Folketal." Accessed 13 October 2020. https://www.dst.dk/da/Statistik/emner/befolkning-og-valg/befolkning-og-befolkningsfremskrivning/folketal.

Stratton, Jon. 2008. "The Difference of Perth Music: A Scene in Cultural and Historical Context." *Continuum: Journal of Media and Cultural Studies* 22: 613–22.

Tregear, Angela, and Sarah Cooper. 2016. "Embeddedness, Social Capital and Learning in Rural Areas: The Case of Producer Cooperatives." *Journal of Rural Studies* 44: 101–10.

Waitt, Gordon, and Chris Gibson. 2009. "Creative Small Cities: Rethinking the Creative Economy in Place." *Urban Studies* 46, nos. 5–6: 1223–46.

Watson, Allan. 2008. "Global Music City: Knowledge and Geographical Proximity in London's Recorded Music Industry." *Area* 40: 12–23.

White, Pauline. 2010. "Creative Industries in a Rural Region: Creative West: The Creative Sector in the Western Region of Ireland." *Creative Industries Journal* 3, no. 1: 79–88.

Wojan, Timothy R., Dayton M. Lambert, and David A. McGranahan. 2007. "The Emergence of Rural Artistic Havens: A First Look." *Agricultural and Resource Economics Review* 36, no. 1: 53–70.

Zasada, Ingo. 2011. "Multifunctional Peri-urban Agriculture: A Review of Societal Demands and the Provision of Goods and Services by Farming." *Land Use Policy* 28, no. 4: 639–48.

Zukin, Sharon. 2010. *Naked City: The Death and Life of Authentic Urban Places*. Oxford: Oxford University Press.

Creative Placemaking Strategies in Smaller Communities

Greg Richards

This final chapter reviews the main themes of the book and contextualizes them within emerging trends of creative tourism, and within the larger field of creative placemaking. This volume brings together an impressive range of analyses of creative tourism in small places around the world. These are based on many different analytical perspectives, including sustainable development, networking, co-creation, producer collaboration, digital technologies, and Indigenous cultures.

Creative tourism is often linked with areas that are in some way disadvantaged in regards to tourism, because it arguably gives places lacking in more traditional tourism resources new opportunities (Duxbury and Richards 2019). For many of these places, small size is a major disadvantage. It often means physical isolation, as in the case of the Azores or the Canadian Arctic, but it also implies a relative lack of the tangible heritage resources that are usually the backbone of traditional cultural tourism.

One of the points made by this volume is that being small can often also be an advantage. For one thing, small places already know they can't engage in the global competitive rat race of city branding and iconic architecture. This is in many ways a good thing, because city branding tends to be superficial, concentrating on just one image or story of a place, blocking all alternative voices for the sake of being "on brand." Although place branding and marketing is where the big money is, there are alternatives for small places that want to put themselves on the map and make themselves better places to live in. The most important of these is "placemaking," which can be viewed as an alternative to place marketing (Richards and Duif 2018). As Hildreth (2009) has pointed out, marketing and branding simply do not work unless the reality of a place matches the image. He suggests, then, that places that want to be successful should improve their reality. The image will follow. If a place is good to live in, it will also be good to visit and to invest in. And places of any size can be good to live in.

This is why there is growing attention nowadays to placemaking. However, while placemaking is in vogue, it is poorly understood. There are many different definitions, and most of these relate to a fairly narrow concept of placemaking as an intervention in the physical environment of a place. But as I will argue here, placemaking is far more than a physical intervention. It is, rather, a complete social practice that involves physical change, as well as changes in thinking and doing.

Placemaking for Smaller, Happier Places

Historically, placemaking has been the preserve of architects and planners. Not surprisingly, they tend to concentrate on the construction of the built environment and how this can improve people's lives. Providing better places to live and work was an ideal of the garden cities movement, and of Jane Jacobs (1961). Many recent scholars have also drawn on Lefebvre's (1991) ideas about the nature of space to show how people interact with the places they inhabit. He argued that physical space was just one element of space as a whole. People also make space through their use of it—the "lived space" of everyday life. And planners and designers and politicians also create ideas about space that influence the way space is viewed and used—"representational space." Physical space, lived space, and representational space form an essential triad, and as such, need to be considered together in the making of places. Most importantly, the spatial triad underlines the essential link between people, communities, and places. People form an attachment to places through living in them, and through using them. Place attachment is also emphasized by Ball (2014), who argues that

> Placemaking is a concept that emerged to describe the intentional process of activating new or existing public spaces to

create that emotional connection. Placemaking, which can take many forms and include a range of activity, activates public space through design, programming, community empowerment, wayfinding, art, marketing—whatever is needed for that particular community. Placemaking is contextual and situational, and whether a project begins with a community's needs or a specific physical location, it will require a unique recipe.

The importance of this emotional connection to place is increasing as people become more mobile. Communities therefore need to find new ways to define themselves. When many people in the developed world can be anywhere they choose, the choice of location is an important one. Why here? Why now? As a result, cities around the world compete not just to be the most powerful or the richest place on earth, but also to be the "place to be."

The phenomenon of global mobility has also produced a new range of explanations for the success of certain places. Perhaps most famously, Richard Florida (2002) has suggested that it is no longer the availability of work that attracts people, but the creativity of places and the presence of other members of the "creative class." Nichols Clark (2003) has also argued that the amenities that places can offer are important in attracting people, including the built infrastructure, cultural facilities, and "atmosphere."

One of the effects of such ideas about the attractiveness of places is that big cities tend to be the favoured locations. These are the world cities, the creative hubs, the storehouses of cultural treasures. Florida (2017) has even suggested that today's major cities are simply not big enough, and that they need to grow bigger to be more efficient and competitive as "superstar cities." At the same time, however, there is a countervailing movement toward smaller places. People fed up with sitting in traffic for hours and paying a fortune for a cramped apartment in the big city have begun to reassess the benefits of metropolitan living. Charles Montgomery's book *Happy City* (2013) rails against ever-increasing urban agglomeration as a process that alienates and makes people unhappy. He advocates smaller-scale developments that bring people into contact with one another, and which make them happier as a result. In Canada, Denis-Jacob (2012,

110) found that the presence of cultural workers is no lower in small cities than in larger ones. Many small places have therefore managed to stop or reverse their previous population decline. Being small, they have managed to foster an emotional connection between people and the place they live in. These are places that many people now want to visit.

In our book *Small Cities with Big Dreams* (2018), Lian Duif and I outline how smaller places can compete with bigger ones if they effectively follow a few basic placemaking principles. The most important of these include having vision and giving that vision meaning for people. People feel connected to places because they have meaning; they are special because of the things that happen there, the experiences we have, and the feelings of identity they create. Our very mobility stimulates us to seek links with and meanings in places. This can take us to the smallest and most peripheral places as well, as the Culture, Sustainability, and Place conference held in the Azores in 2017 illustrated. In bringing the conference to the Azores, Nancy Duxbury drew on her extensive international networks to make this small archipelago a focus of global attention for a short while. One of the emerging themes of the discussions at the conference was the way in which the Azores have been at the forefront of globalization for the past five centuries, as part of an essential survival strategy. Small places such as the Azores have been influenced by, and have in turn influenced, globalization just as large urban centres have done. Smaller places and peripheral locations face similar challenges as a result of global changes, but they have a smaller resource base with which to confront these challenges. Networks, along with the ability to connect with other places and mobilize people at a distance, are an essential key to extending this resource base. Small places, as Nancy Duxbury points out in chapter 1 with her analysis of the situation in Portugal, often suffer from a lack of ambition, or a lack of belief in their own capabilities. The CREATOUR project in non-metropolitan places in Portugal is one concrete attempt to address this credibility gap.

Making the necessary connections between the local and global and between tourists in search of meaning and the local meanings of culture, takes a lot of imagination, a lot of creativity (Richards, Wisansing, and Paschinger 2018). We need to be creative to understand what resources and meanings we can offer to the inquisitive tourist or the global investor.

The creativity of places lies not just in the formal types of creative industries that are now so popular with governments seeking to stimulate economic development. Creativity also lies in the everyday life of places, in the daily rhythms of work and play, in the ingenious ways in which people have adapted to the world around them. Usually these things are almost invisible to the "locals," like water is to the fish that swim in it. In order to frame their creativity for others, places must therefore first think creatively about what they have and how this could be interesting for others.

These issues are also considered by Richards and Marques (2012), who note that

> "Creative tourism is a form of networked tourism, which depends on the ability of producers and consumers to relate to each other and to generate value from their encounters. Creative tourists are 'cool hunters' in search of creative 'hot-spots' where their own creativity can feed and be fed by the creativity of those they visit" (10).

The creative tourist is therefore seeking interesting encounters with local people and their creativity, which they can then turn into creative capital that will increase their knowledge, skills, and creative status. A number of recent studies have provided profiles of these mobile consumers. For example, Tan, Luh, and Kung (2014) confirmed that creative tourists are novelty-seekers who want to acquire knowledge and skills and who have an awareness of environmental issues. Huang, Chang, and Backman (2019) also found creative tourists to be predominantly female, relatively young (thirty-one to forty being the largest age group), well-educated, and with a relatively high income. This profile is fairly close to that of the cultural tourist, with the main distinguishing feature being the motivation to engage in active creative experiences.

This desire for active engagement is an important potential asset for placemaking. Placemaking can be seen as a combination of three essential elements: resources, meaning, and creativity (Richards and Duif 2018). The resources of a place can more effectively be exploited by giving them meaning for those within and beyond small places, which in turn requires creative thinking and action. This triad of elements deliberately mirrors the three elements of social practices identified by Shove, Pantzar, and Watson (2012): materials, meaning, and competences. To use a suitable creative example, the practice of painting embraces all of these elements. A painter needs basic resources such as paint and canvas, as well as resources that might trigger creative ideas, such as landscapes or cityscapes. But having these resources available is not enough: they also need to be used in a competent way that gives meaning to the finished work of art, both for the artist and the audience.

Placemaking is also an art, particularly if we consider what Markusen and Gadwa (2010) have termed "creative placemaking." The problem is that too many places have bought into the rhetoric of "creative cities" and the "creative industries," seeing artists, architects, and other members of the "creative class" as central to the process. Although there are members of the creative class everywhere, even in the most remote locations (Brouder, 2012), most places simply do not have the concentration of artists that will feed a creative cluster or a biennale. For analysts of the creative class and creative industries like Richard Florida and Alan Scott, only big cities can really claim to be creative hubs.

But this ignores the everyday creativity that is present in all places and that visitors also increasingly want to experience. The everyday creativity that is embedded in local lifestyles is what makes most places. It is found in crafts, pastimes, the arts, music, and literature. These are among the aspects of creativity that help to keep places distinctive in a globalizing world. These were also the types of creativity that inspired the original concept of "creative tourism," which Raymond and I originally defined as

> Tourism which offers visitors the opportunity to develop their creative potential through active participation in learning experiences which are characteristic of the holiday destination where they are undertaken. (2000, 18)

This definition was based on the idea that tourists could develop a relationship with the places they visited through learning about local creativity. Originally, we thought about this in terms of formal learning: courses or workshops. But as we gained more experience with the concept, we realized that most people don't want to spend all of their holiday in the classroom or atelier. What most people want is an experience, a taste of creativity that will enable them to develop their own knowledge and skills, as well as provide a relationship with the people they were visiting. The most memorable part of a cookery class is usually not the recipe or even the food, but the people who were sharing their creative skills. This was when we began to see that creative tourism was not just about learning, but also about creating relationships. Creative tourism seems to work particularly well in the "lived space" of the everyday, where people can encounter one another on an equal footing. Creative tourism can therefore also be a form of "relational tourism" (Richards 2014).

The relationships formed via creative tourism also tend to be of a particular type. People seek out the "local" creative, often their equal in terms of knowledge and skills, but embedded in a different local context. We could argue that the local has become the new touchstone of authenticity or originality (Russo and Richards 2016). We want to go where the locals go, do what they do, experience what they experience. As Kathleen Scherf shows in her introduction, the desire for the local is now being made tangible in the creation of "localhoods" in the city of Copenhagen (Richards and Marques 2018). The "live like a local" phenomenon is now widespread on the Internet via sites such as Spotted by Locals. Even global companies such as Airbnb offer their clients the chance to "belong anywhere." But the reality of this kind of "relationship" often entails a brief encounter with a gentrifying property developer or one of their staff, who hands over the keys and a guide to local restaurants before heading to the next client.

Much more sustainable portals to the local are provided by creative links. People who want to share their creativity and skills with each other are likely to form more lasting relationships than hotel staff and their guests. Learning a skill involves extensive face to face contact with those who have the skills we are seeking. The focus on skills also removes the problem of the language barrier that usually restricts entry to the tourist market. Sharing interest in skill is a great leveller—it can reduce barriers of gender, class, and origin. The common (often non-verbal) language of making and doing is an important form of communication.

Creative skills and knowledge are also widely present in the host community. People are usually looking not for "experts" but for people with whom they can share an emotional link to a creative process and, by extension, the place they are in. In this sense, the act of "doing" is key. Crispin Raymond summarized the essence of creative tourism with a saying from Confucius: "I hear and I forget. I see and I remember. I do and I understand." It is the doing that is essential in social practices (Baerenholdt 2017). By doing things together we share information, ideas, and feelings. This is very different from the normal range of interactions between tourists and locals, which tend to be acted out in a scripted way at reception desks or restaurant tables.

Creative tourism as a way of sharing experiences and skills between tourists and locals moves the focus of tourism into the sphere of daily life. This tends to enrich the tourist experience, but it also increases the vulnerability of local culture. When sharing the creative process, it is much more difficult to confine tourists to the "front stage"—they want a backstage pass. But you have to be confident that those who share an enthusiasm for creativity will be relatively small in number and positive in demeanour.

Placemaking through Creative Tourism

If we view placemaking as a practice that unites the elements of resources, meaning, and creativity, then we can start to chart the potential contributions of creative tourism to this process.

In terms of resources, creative tourism can be a means of conserving those things that the local community might otherwise be in danger of losing. Just as cultural tourism has been important in stimulating the conservation of tangible heritage in recent decades (UNWTO 2018), now creative tourism is providing more possibilities for conserving intangible heritage as well. Very often the everyday creative skills of the community are losing ground to new areas of creativity that are more attractive to young people. One way to get young people interested in the creative legacy of the community is to show that it is important and that it can generate more resources and opportunities. By introducing tourists to the creative products and processes of a community, these can be valorized and successive generations can find new ways of interacting with them. This was one of the original inspirations for the creative tourism concept— the rediscovery of traditional craft skills through the EUROTEX project in the Alto Minho region of Portugal, on the island of Crete in Greece, and in Finnish Lapland (Richards 1999). By taking traditional crafts and retrofitting them to be attractive to new generations, the pool of craft producers and the potential tourism market could be expanded as well.

In terms of meaning, the development of creative tourism generated new ways of looking at the relationship between communities and the areas they live in. Through the EUROTEX project, textile crafts suddenly became a source of income in rural areas. More importantly, the fact that tourists were coming to learn traditional skills changed the meaning of the tourist-local relationship from host and guest, server and served, into a relationship of equals—people interested in the same skills and creative processes. Creative tourism also became attached to new meanings as an alternative to "mass cultural tourism" and as a touchstone for authenticity (Richards 2018).

This led to the realization that creative tourism could be an important path for placemaking, and that the use of tourism to identify, concentrate, and harness creativity was an important potential means for making places better. If we view creative tourism as a practice, as a means of doing, then the power of creative tourism compared to more conventional forms of tourism becomes more evident. When we examine the practice of creative tourism more closely, we also begin to identify the essential elements of practice that make it different. Going back to our triad of placemaking, the action of developing creative tourism also falls into three basic areas:

1. "what we have," including materials, people, and the knowledge endowments of a place;

2. "how we do things," which encompasses the application of creativity to the use of our resources, including governance modes and the representational nature of place; and

3. "what we do" to implement knowledge- and creativity-related policies and projects.

For many small places there is also a need to do things differently, particularly in relation to larger places. As Comunian and England (2018) argue, smaller places will almost inevitably have to rely on bottom-up strategies that make effective use of the relatively limited resources they have. In the development of creative tourism, small places need to employ a variety of different strategies to overcome their relative disadvantage.

In the wake of COVID-19, creativity will also be essential to helping smaller places recover. Although it might be expected that flight from overcrowded cities will help smaller places attract tourists again, they will have to convince people that it is safe to come. The personal relationships built up through creative development may be a key factor in this.

Lessons Learned About Creative Tourism in Smaller Communities

The variety of different cases examined in this book underline that places are not just sites for creative activity: they are also an important context that shapes, and is also shaped by, creativity. Place-related creativity stands in contrast to the more globalized models of creative development that have been propagated by researchers such as Richard Florida. In small places in particular, the context of creativity is key. Unlike larger cities, where cosmopolitan forms of creativity circulate freely and the creative class provides a ready market, small places have to use their endogenous creativity to survive. In smaller communities, place itself becomes an important asset because often it is one of the few resources available. The centrality of place also provides an important link to the concept of placemaking.

Among the lessons emerging from the cases presented in this volume is the strong role played by local communities and social networks in developing and supporting creativity. Placemaking can be seen as a collective effort by individuals/groups to reimagine and remake the environments that surround them (Strydom, Puren, and Drewes 2018). Placemaking processes have been propelled to the foreground by the desire of local communities for more control over the places they live in, and their growing concern with quality of life, well-being, and conservation.

In a globalizing world, the role of place becomes more important than ever. When you can choose to be anywhere, where you choose to be matters. Where you spend your time matters, who you are with matters, when you are there matters. Just by making such choices, you are already involved in the placemaking process. This is why the renewed attention to smaller places is so important. A meaningful implication is that instead of writing small places off as economically uncompetitive, many people are choosing to be in smaller places because of the other qualities they offer: a human scale, social cohesion, and a slower pace of life (Richards and Duif 2018).

We have also learned that taking advantage of the creative energy of local communities means paying attention to the factors that enable creativity and creative placemaking. The cases in this volume emphasize the importance of governance mechanisms, particularly in terms of facilitating bottom-up processes of decision-making and creativity. This will often mean empowering networks in small places, particularly where those networks can link to the outside world and provide new opportunities. In this sense, tourists can generate the weak links that bridge to other places and carry new ideas and ways of doing.

How can such links be made between locals and tourists? In the past this usually required some kind of physical intermediary, but now platforms such as Airbnb are offering peer-to-peer connections, also for creative experiences. The question is whether this will produce new forms of interaction or simply become a new way of distributing experiences? As the market for destination experiences becomes more consolidated, there is a danger that these will also become more standardized. If, however, new intermediaries and producers are facilitated by the possibilities offered by new technologies, then there should be more opportunities for innovation and creative development.

Directing the relationships between locals and visitors has also now become a task for the public sector in many places, as many of the chapters in this volume illustrate. The recent report by United Cities and Local Governments highlights the importance of giving both tourists and locals a role in the social, cultural, and creative life of places (Richards and Marques 2018). The interface between the "global nomad" (Richards 2015) and the places they visit, dwell, or stay in is arguably by its very nature creative, since it involves a constant negotiation and renegotiation of identity and belonging.

In such negotiations, giving tourists a stake in the places they visit is important, as Barcelona has recognized in its efforts to designate tourists as "temporary citizens." The temporary citizen concept is of course a formalistic way of saying to visitors that they can share the city with residents, enjoying the same rights, but also shouldering the same duties with respect to the use of public space and consideration for others. The less formal and more effective means of belonging lies in the relationality of the place itself, in fact; as Randall Collins (2004) argues, "shared co-presence" and a common focus of attention creates an "emotional energy" that makes us feel good about ourselves and about being in a place. This emotional energy can already be attained by passive participation in rituals such as visiting the Louvre to view the *Mona Lisa*. But it can be much greater in situations where we are actively involved in the creative process, particularly when we are connected to others who can share with us the skills, meanings, and understandings attached to the place we are in. This is the type of energy that can lead to very effective creative placemaking.

The contributions to this volume also show us that one of the key goals is to get "on the map" and attract attention—which is an increasing challenge in the face of the concentration of media power in major cities. Brabazon (2014) outlines the plight of smaller, slower places in the competitive race against "fast" global cities. She sees the neo-liberal rhetoric of the creative class and cool cities as "corrosive" to small cities. As in many of the cases presented in this volume, she sees the context of place as being vital to understanding the dynamics of small cities. Cities that often flourished in the industrial age are floundering in the information age, often because they are not effectively networked, either externally or internally. To build such links, places need to find a purpose. Brabazon argues that events are often central to finding a purpose and aligning stakeholders behind a development agenda. There are opportunities for smaller places (or "third-tier cities") to grab attention if they can organize the right kind of programs and events. This has been illustrated in many of the contributions to this volume, which show how events can act as a focal point for attracting people, attention, and resources, and giving new purpose to small places.

Conclusions

In a rapidly globalizing world, the serial reproduction of culture is giving new urgency to the distinction of places. While most places still seek distinction in place marketing and branding, a growing number of communities are realizing that there are other routes. A number of these are outlined in this volume. Above all, the importance of harnessing the creative energy of local communities is an important factor.

Most importantly, places need to improve their reality, not their image. If the reality is attractive, the image will be too. These realities can be improved through the provision of physical infrastructure, but far more important these days is the development of "soft infrastructure," and in particular the creative skills needed to link the local to the global. Tourism is one important route for doing this, but the flow of people, ideas, and resources has to be locally controlled. It needs to be developed by people who are embedded in the places they come from, and who can help interpret the creative resources of that place to others. This act of sharing can help to stimulate the emotional energy that is needed to drive the placemaking process. Creative tourism is an effective gateway to developing shared experiences between "tourists" and "locals" and ensuring that the creativity of placemaking is maintained.

References

Baerenholdt, Jørgen. 2017. "Moving to Meet and Make: Rethinking Creativity in Making Things Take Place." In *The SAGE Handbook of New Urban Studies*, edited by John Hannigan and Greg Richards, 330–41. London: SAGE. http://dx.doi.org/10.4135/9781412912655.n21.

Ball, Richard. 2014. "Economic Development: It's About Placemaking." *European Business Review*, 19 September 2014.

Brabazon, Tara. 2014. "Go to Darwin and Starve, Ya Bastard: Theorizing the Decline of Third Tier Cities." *Fast Capitalism* 11, no. 1. http://www.uta.edu/huma/agger/fastcapitalism/11_1/brabazon11_1.html.

Brouder, Patrick. 2012. "Creative Outposts: Tourism's Place in Rural Innovation." *Tourism Planning & Development* 9, no. 4: 383–96.

Collins, R. 2004. *Interaction Ritual Chains*. Princeton and Oxford: Princeton University Press.

Comunian, Roberta, and Lauren England. 2018. "Creative Regions: From Creative Place-Making to Creative Human Capital." In *Handbook on the Geographies of Regions and Territories*, edited by Anssi Paasi, John Harrison, and Martin Jones, 169–81. Cheltenham, UK: Edward Elgar.

Denis-Jacob, Jonathan. "2012 Cultural Industries in Small-Sized Canadian Cities: Dream or Reality?" *Urban Studies* 49, no. 1: 97–114.

Duxbury, Nancy, and Greg Richards, eds. 2019. *A Research Agenda for Creative Tourism*. Cheltenham, UK: Edward Elgar.

Florida, Richard. 2002. *The Rise of the Creative Class: And How It's Transforming Work, Leisure, Community and Everyday Life*. New York: Basic Books.

———. 2017. "Why America's Richest Cities Keep Getting Richer." *The Atlantic*, 12 April. https://www.theatlantic.com/business/archive/2017/04/richard-florida-winner-take-all-new-urban-crisis/522630/.

Hildreth, Jeremy. 2008. *The Saffron European City Brand Barometer: Revealing which Cities Get the Brands They Deserve*. London: Saffron Brand Consultants. http://directe.larepublica.cat/documents/noticies/saff_citybrandbarom.pdf.

Huang, Yu-Chih, Lan Lan Chang, and Kenneth F. Backman. 2019. "Detecting Common Method Bias in Predicting Creative Tourists Behavioural Intention with an Illustration of Theory of Planned Behaviour." *Current Issues in Tourism* 22: 307–29.

Jacobs, Jane. 1961. *The Death and Life of Great American Cities*. New York: Vintage.

Lefebvre, Henri. 1991. *The Production of Space*. Translated by Donald Nicholson-Smith. Oxford: Basil Blackwell.

Markusen, Ann, and Anne Gadwa. 2010. "Creative Placemaking." Mayors' Institute on City Design and the National Endowment for the Arts. https://www.idsa.org/sites/default/files/CreativePlacemaking-Paper.pdf.

Miettinen, Satu, Jaana Erkkilä-Hill, Salla-Mari Koistinen, Timo Jokela, and Mirja Hiltunen. 2019. "Stories of Design, Snow, and Silence: Creative Tourism Landscape in Lapland." In *A Research Agenda for Creative Tourism*, edited by Nancy Duxbury and Greg Richards, 69–84. Cheltenham, UK: Edward Elgar.

Montgomery, Charles. 2013. *Happy City: Transforming Our Lives through Urban Design*. London: Macmillan.

Nichols Clark, Terry. 2003. "Urban Amenities: Lakes, Opera, and Juice Bars: Do They Drive Development?" In *The City as an Entertainment Machine*, edited by Terry Nichols Clark, 103–40. Bingley, UK: Emerald Group Publishing.

Richards, G., Wisansing, J. and Paschinger, E. 2018. *Creating Creative Tourism Toolkit*. Bangkok: DASTA.

Richards, Greg. 1999. *Developing and Marketing Crafts Tourism*. Tilburg, NL: ATLAS.

———. 2014. "Creating Relational Tourism through Exchange: The Maltese Experience." *Journal of Hospitality and Tourism* 12, no. 1: 87–94.

———. 2015. "The New Global Nomads: Youth Travel in a Globalizing World." *Tourism Recreation Research* 40, no. 3: 340–52.

Richards, Greg, and Lian Duif. 2018. *Small Cities with Big Dreams: Creative Placemaking and Branding Strategies*. London: Routledge.

Richards, Greg, and Lénia Marques. 2012. "Exploring Creative Tourism: Introduction." *Journal of Tourism Consumption and Practice* 4, no. 2: 1–11.

Richards, Greg, and Crispin Raymond. 2000. "Creative Tourism." *ATLAS News* 23: 16–20.

Richards, Greg, and Julie Wilson. 2006. "Developing Creativity in Tourist Experiences: A Solution to the Serial Reproduction of Culture?" *Tourism Management* 27, no. 6: 1209–23.

Russo, Antonio, and Greg Richards. 2016. *Reinventing the Local in Tourism: Producing, Consuming and Negotiating Place*. Bristol, UK: Channel View Publications.

Shove, Elizabeth, Mika Pantzar, and Matt Watson. 2012. *The Dynamics of Social Practice: Everyday Life and How It Changes*. London: SAGE.

Strydom, Wessel, Karen Puren, and Ernst Drewes. 2018. "Exploring Theoretical Trends in Placemaking: Towards New Perspectives in Spatial Planning." *Journal of Place Management and Development* 11, no. 2: 165–80.

Tan, Siow-Kian, Ding-Bang Luh, and Shiann-Far Kung. 2014. "A Taxonomy of Creative Tourists in Creative Tourism." *Tourism Management* 42: 248–59.

United Nations World Tourism Organization (UNWTO). 2017. *Tourism and Culture Synergies*. Madrid: UNWTO.

CONTRIBUTORS

JON ANDERSON is professor of human geography at the School of Geography and Planning, Cardiff University, United Kingdom. His research interests focus on the relations between culture, place, and identity, particularly the geographies, politics, and practices that emerge from these. His key publications include *Understanding Cultural Geography: Places and Traces* (Routledge, 2010, 2015 [2nd ed.]), *Water Worlds: Human Geographies of the Ocean*, co-edited with Kimberley Peters (Routledge, 2014), and *Page and Place: Ongoing Compositions of Plot* (Brill Rodopi, 2014).

GEOFF ANTROBUS is professor emeritus in economics, Rhodes University, South Africa (MSc, Natal; PhD, Rhodes), and senior research fellow with the South African Cultural Observatory, focusing on festivals and events. His other economic impact studies include those on the the conversion of livestock farms to private game reserves, on foreign students studying at Rhodes University, on livestock theft, and of the potential removal of the High Court from the city of Grahamstown.

JESSICA FAUSTINI AQUINO, PhD, and is assistant professor at Hólar University and head of the Tourism Research Department, Icelandic Seal Center. Her research interests are in tourism experience from the perspective of residents and tourists; the potential impact that tourism has on community development and responsible management of natural areas; and sustainable tourism and responsible practices in Arctic coastal communities and seascapes.

SUZANNE DE LA BARRE, PhD, teaches in the Department of Recreation and Tourism Management and the Master of Arts in Sustainable Leisure Management Program at Vancouver Island University, in British Columbia, Canada, and the University of the Arctic's Graduate Certificate in Northern Tourism. Her research focusses on the circumpolar world, economic diversification and community development, and the creative and cultural industries. A long-time Yukoner, she lives in the territory's capital, Whitehorse, and Nanaimo, British Columbia.

GEORGETTE LEAH BURNS, PhD, is senior lecturer and environmental anthropologist in the Environmental Futures Research Institute and School of Environment and Science at Griffith University in Brisbane, Australia. The former head of both the Rural Tourism Department at Hólar University College and Tourism Research at the Icelandic Seal Centre, her research focuses on understanding and effectively managing the interactions between people and wildlife in tourism settings.

FIONA DRUMMOND obtained her bachelor's and honours with distinction, majoring in geography and economics, from Rhodes University in South Africa. Her honours research project on cultural and creative mapping in rural areas won the Economics Society of South Africa's Founder's Prize for the best honours research project in the country. She is currently finalizing her master's on CCI clustering and its link to development in non-metropolitan areas.

JAMES DRUMMOND is senior lecturer in the Department of Geography at North-West University, in Mahikeng, South Africa, where he has taught for over thirty years. He has degrees from the Universities of Glasgow in Scotland and Witwatersrand in South Africa, and has published on rural and agricultural development in South Africa. His current interests lie in small-town development and the cultural and creative economies.

NANCY DUXBURY, PhD, is senior researcher at the Centre for Social Studies, University of Coimbra, and a member of the European Expert Network on Culture. She is the principal investigator of CREATOUR. Her research has examined culture in local sustainable development; culture-based development strategies in smaller communities; and cultural mapping, which bridges academic inquiry, community practice, and artistic approaches to understand and articulate place. Her recent books include *Animation of Public Space through the Arts: Toward More Sustainable Communities* (Almedina, 2013), *Cultural Mapping as Cultural Inquiry* and

Culture and Sustainability in European Cities: Imagining Europolis (both Routledge, 2015), *Artistic Approaches to Cultural Mapping: Activating Imaginaries and Means of Knowing* (Routledge, 2018), and *A Research Agenda for Creative Tourism* (Edward Elgar, 2019).

DIMITRI IOANNIDES is professor of human geography and director of the European Tourism Research Institute (ETOUR) at Mid-Sweden University. He has written extensively on matters relating to the economic geography of tourism and on tourism in the context of sustainable development. He is a series editor of Routledge's New Directions in Tourism Analysis series.

M. SHARON JEANNOTTE is senior fellow at the Centre on Governance, University of Ottawa. She has published research on a variety of subjects, including the points of intersection between cultural policy and social cohesion, the role of culture in building sustainable communities, culture and volunteering, immigration and cultural citizenship, cultural mapping as a tool for place-making, cultural indicators at the local level, and provincial/territorial cultural policy and administration in Canada.

JEFFREY MORGAN is a software developer with research interests in data visualization, geographical information systems, and human-computer interaction.

EVANGELIA PETRIDOU is assistant professor of public administration at the Risk and Crisis Research Center and Mid-Sweden University. She is a public policy and public administration scholar. Her recent work has appeared in *Policy Studies Journal* (2014), *Policy and Society* (2017), and *Policy Studies* (2018), and she was co-editor, with Inga Narbutaité Aflaki Lee Miles, of *Entrepreneurship in the Polis: Understanding Political Entrepreneurship* (Ashgate, 2015).

SOLÈNE PRINCE recently completed her PhD in tourism studies at Mid-Sweden University, where she works as a researcher at the European Tourism Research Institute (ETOUR). Her doctoral research led her to explore the livelihoods of creative actors in peripheral Denmark. She is currently a post-doctoral researcher at Linnaeus University, Sweden. Some of her latest work has appeared in the *Journal of Sustainable Tourism* (2017), *Landscape Research* (2018) and *Tourist Studies* (2018).

GREG RICHARDS is professor of place-making and events at Breda University and professor of leisure studies at Tilburg University in the Netherlands. He has published widely in the areas of cultural and creative tourism.

KATHLEEN SCHERF is professor of communication at Thompson Rivers University in British Columbia, Canada. She earned a PhD from the University of British Columbia. Her main areas of research are resident-tourist relationships and how that dynamic impacts perceptions and experiences of local culture. She is interested in creative tourism, cultural mapping, and deep mapping viewed through a communication lens. She is the editor of this volume.

SUSAN L. SLOCUM is associate professor in the Department of Tourism and Events Management at George Mason University, Manassas, Virginia. Sue has worked on regional planning and development for fifteen years and worked with rural communities in Tanzania, the United Kingdom, Belarus, and the United States.

KIERON SMITH is a research associate on the *Literary Atlas* project at Cardiff University. He completed his PhD, a study of the documentary films and poetry of John Ormond, at Swansea University in 2014. His current research focuses on the conceptual and creative potentialities of digital cultural geographies.

JEN SNOWBALL is professor of economics at Rhodes University, South Africa, where she earned a PhD in economics. She is also a researcher at the South African Cultural Observatory, a research organization focused on the cultural and creative industries in South Africa funded by the national Department of Arts and Culture. Her research interests are focused mainly in the field of cultural economics, specifically the use of market and non-market valuation methods, primarily as they apply to cultural festivals. Her recent work for the Cultural Observatory has been on cultural mapping studies, employment in the cultural and creative industries, and developing a framework for the monitoring and evaluation of publically funded arts, culture, and heritage.

EUGENE THOMLINSON is associate professor and director of the School of Tourism and Hospitality Management and head of the Bachelor of Arts in Global Tourism Management Program at Royal Roads University. He has approximately twenty years of tourism and hospitality experience, both in industry and academia. His research interests and experiences focus on screen tourism, sustainable development, experiential marketing, interpretation, persuasive communication, image and branding, and mega-event sports tourism (e.g., the Commonwealth Games, Olympic Games, World Cup, etc.). He also enjoys working with communities and organizations on tourism development plans and opportunity assessments.

ELISABETE TOMAZ is a researcher at DINAMIA, ISCTE-IUL (Research Group Cities and Territories). She completed a PhD in sociology at ISCTE-IUL, a project funded by FCT (the Portuguese public agency that supports science, technology, and innovation). At the same time, she participated in the COST ACTION IS 1007 "Investigating Cultural Sustainability" project and collaborated with INTELI (a centre of innovation in Lisbon) on several European and municipal councils' projects in territorial development, urban planning, and innovation policies. She holds a bachelor's degree in communication design (Faculty of Fine Arts, University of Lisbon) and a master's in communication sciences (Faculty of Human Sciences, Portuguese Catholic University).

CHRISTINE VAN WINKLE is committed to community-based research examining visitor experiences at events and attractions. As a former festival coordinator and attraction consultant, Dr. Van Winkle brings both practical experience and theory-based research to inform her practice. Dr. Van Winkle's work has been published widely and appears in a range of tourism, leisure, and event journals, books, conference proceedings, and reports. She is professor in the Faculty of Kinesiology and Recreation Management at the University of Manitoba.

INDEX

A

Aalto, Alvar, 69–70
Aarø-Hansen, Mikkel, 7
ACAB. *See* Arts and Crafts Association Bornholm
AECO (Association of Arctic Expedition Cruise Operators), 152, 154
agriculture, 192–193, 194, 197, 202
Airbnb, 289, 293
Aitchison, Cara, 194
Albuquerque, New Mexico, 243
Allianz, 146
Allinson, Johanne, 193
Álvarez-García, José, 8
Amundsen, Roald, 143
Andersen, Lisa, 115, 259, 266
Anderson, Jon, 209, 219, 221, 224, 225
Anselmi, Elaine, 149
Antrobus, Geoff, 79, 85, 89, 90
Aquino, Jessica F., 81, 138, 165, 168, 171, 176, 181, 185, 198, 211
archaeological experiences, 67–68
Arctic Air (television), 235
Arctic Bay Adventures Co-op, 156
Arctic regions, 109, 110–111, 116, 128–129, 143–144, 153
    See also Northwest Passage
Areas of Archaeological Importance, 67
Army Wives (film), 245
Arnarsdóttir, Eygló Svala, 175, 176
artists
    clusters and networks, 264–265, 268–269, 276–278
    expansion of repertoire, 150–151
    professional distinctiveness, 270–271
    in remote and smaller places, 44, 115, 126–127
    values and motives, 265
arts
    and authenticity, 150
    placemaking as, 288
residents' engagement with, 272, 273–276
    in rural Iceland, 170
    tourists' spending on, 147–148
    value indicators, 91–92
    in Yukon and Canada's North, 123–127, 129, 140
arts and crafts
    banned from imports, 147–148
    CREATOUR pilots, 40
Arts and Crafts Association Bornholm (ACAB)
    case study approach, 267
    membership and funding, 268–269
    operations, 260, 269–271, 277
Ashworth, Gregory, 66
Association of Arctic Expedition Cruise Operators (AECO), 152, 154
Atterton, Jane, 264
Australia and creative peripheries, 114–115
authenticity
    Arctic arts and crafts, 150
    vs. distinctiveness, 213–214
    everyday experiences, 151, 289
    historical experiences, 65, 66
    place-specific, 82–83, 92, 115, 263
    tourists' desire for, 31
Azizi, Saleh, 197, 202, 203
Azores, 286

B

Backman, Kenneth F., 287
backpackers, 199
Baerenholdt, Jørgen, 290
Bakas, Fiona Eva, 38, 39
Ball, Richard, 284
Bantustans, 86–87
Barcelona, 7–8, 293
Baroque Nights, 65
Bates, Charlotte, 221
Batswana culture preservation, 86–87, 88, 90
Baxter, Pamela, 119

Beaumaris, Wales, 223–225
Beaumont, Narelle, 278
Beeton, Sue, 243, 244, 245, 246, 247, 249, 250, 252
Bell, David, 6, 260, 262, 263, 278
Bellini, Nicola, 62
Bennett, Dawn, 265
Bennett, Nathan, 140
Berg, Bruce, 231
Bessière, Jacinthe, 199, 203
Big River Analytics, 140, 147
Binette, Sylvie, 121
Binkhorst, Esther, 191, 198, 202
Bjarnadottir, Erna, 175
Blain, Carmen, 244
Blapp, Manuela, 28, 195, 204
Bohm, Steffen, 91
Bolan, Peter, 230
Bolwig, Simon, 193
Bonifacio, Glenda, 5
Borba, Carla, 195, 202, 203, 204
Borðeyri, Iceland, 167, 168
Borgatti, Stephen P., 264
Borkman, Thomasina, 192, 202
Bornholm, Denmark, 260, 267–268
Bourdieu, Pierre, 209
Boyd, Stephen W., 115
Brabazon, Tara, 294
Bräcke, Sweden, 273–274
Bramham, Daphne, 148
branding, 177, 197, 200, 213, 283–284
Breaking Bad (television), 243
Brouder, Patrick, 8, 17, 34, 45, 110, 288
Brown, Chris, 147
Brown, Frances, 114, 115
Brundtland Report (*Our Common Future*), 9
Buffalo Airways, 246
Bunten, Alexis, 112
Burns, Georgette L., 81, 138, 165, 166, 169, 171, 175, 198, 211
businesses
    artistic, 265
    in CREATOUR pilots, 38–40, 42
    entrepreneurial, 71, 115, 183, 196–197, 198, 202–203
Butler, Richard, 139

C

Camagni, Roberto, 266
Campbell, Heather, 81

Canada's North
    cruise tourism impacts, 137–138, 139, 145–151
    cultural capital and community preservation, 140
    governance of, 151–152, 153–154, 156
    populations and economies, 141–143
    shipping and cruise tourism evolution, 143–145
    sovereignty, 144–145
    See also Nunavut, Canada; Yukon, Canada
Canadian Coast Guard, 144, 146
Canadian populations, 5
Cape Dorset, Nunavut, 147, 150
capital
    creative, 112, 115, 118
    cultural, 140, 156–157
Cardoso, Lucilia, 229, 236, 241, 250
Carl, Daniela, 230
Carothers, Courtney, 165
Carson, Doris, 116, 118
Cartographies of Culture (Damian Walford Davies), 218
Carvalho, Claudia Pato de, 38
case study research, 119, 231–232
Castro, Tiago Vinagre de, 30, 34, 36, 39, 41, 42
Cavaliere, Christina, 195, 201, 203
Cavin, Philip, 110, 114
Cawley, Mary, 165
CBC (Canadian Broadcasting Corporation), 235
celebrity endorsements, 243–244
Český Krumlov District, Czech Republic, 64–66
Český Krumlov State Castle, 65
Chambers, Catherine, 165, 171
Chang, Lan Lan, 287
cheese-making, 175, 197
Chen, Chien-Yu, 230, 242, 244
Chernobyl (television), 243
Cheung, Lewis T. O., 195, 200, 203
China, festivals, 81
circumpolar region. *See* Arctic regions
cities. *See* communities; smaller communities; urban centres
City of Quarters: Urban Villages in the Contemporary City (David Bell and Mark Jayne), 6
City of the Unexpected event, 214
Clifton, Nick, 110
climate change

impacts to Arctic life, 146
and tourism, 138, 145
Cloke, Paul, 110, 259, 260, 263
Clough, Greta, 182
clusters
    ACAB, 268–269
    CREATOUR project, 35
    Jyväskylä, 69
    York, 68
    See also creative clusters; networks
Cole, Alexander, 259, 264
Collins, Patrick, 44, 110, 114, 118
Collins, Randall, 294
Columbia tourism, 243
commodifications, 28, 82, 111, 112, 244
communities
    creative, 181–182, 239–240, 243–244, 285
    development and economy, 74–75, 92, 110, 260
    as film locations, 239, 244–246, 251–252
    identity, 185, 195
    and placemaking, 284–285
    pride, 175–176, 194
    sustainability of, 211
    tourism in, 148–149, 150, 156–157
    younger generations, 170, 179–180, 290–291
    See also creative peripheries; Indigenous communities; residents; smaller communities; visitor-resident interactions
community engagement
    in CREATOUR pilots, 31, 41, 44–45
    Húnaþing vestra initiatives, 168–169
    networking and support, 39–40, 44–45, 102, 182
    planning and development, 9–11, 184–185, 230, 250–251, 252, 254
    skills-sharing, 289–290
    See also public input
Comunian, Roberta, 264, 265, 291
Conference Board of Canada, 142
conferences, 11, 36, 271, 286
Connell, John, 263
consumerism, 209
Cooper, Sarah, 264
Copenhagen "Localhood" campaign, 7
Cornwall, England, 263
Couret, Caroline, 34
Cranston, Bryan, 235
creative capital, 112, 115, 118

Creative City Network of Canada, 183, 253
creative clusters, 253–254, 259–260, 262, 265–266, 288
creative economy, and tourism, 194
creative industries
    definitions and scope, 182, 230, 261–262
    impact studies and research, 110, 112–113
    and place, 195
    rural, 259–260, 262–263
    tourism as, 165
    in urban development, 63
Creative Óbidos, 71
creative peripheries, 114–116, 118
creative spaces, 260, 270–271
creative tourism
    as conservation, 290–291
    definitions and trends, 4–5, 28, 29, 229–230, 288–289
    as experiences, 111, 140, 174–175, 287
    filmmaking contributions to, 239–241, 249–250
    and identity, 191–192, 195
    and placemaking, 291–292
    as relational tourism, 289–290
    as side businesses, 42
    in smaller places, 43–44, 210
    sustainable development, 8–9, 20–22
creative tourism research
    Mahika Mahikeng Festival study, 91–97
    state of, 8–9, 44, 61, 111, 118
    urban bias, 39
    See also under Yukon, Canada
creativity
    defined, 261
    elements, 202, 230–231
    place-related, 292
    as resilience, 165, 176
    tourists' self-expression, 29–30, 287–288
CREATOUR project
    components, 35–37
    goals and design, 28–31, 43–45, 286
    knowledge sharing, 9, 38
    pilot observations, 38–42
    pilot selection, 34, 37–38
    pilots and projects, 47–59
    structure and coordination, 31–34
Creswell, John W., 267
Croce, Erica, 199, 201
Crompton, John L., 85
Cronon, William, 115
Croy, W. Glen, 229, 230, 244, 249, 251, 254

cruise tourism
  in Canada's North, 144–145
  impacts to communities, 148–149, 150, 153
  regulation, 152, 154–155
  and sustainability, 139, 155–156
Crystal Serenity, 5, 137, 148, 149
Cudny, Waldemar, 81, 84, 85
cultural activities
  definitions, 111
  impact valuation, 91–92
  See also events; festivals
cultural capital, 140, 156–157
cultural mapping, 10–11, 17, 226, 253–254
Cultural Mapping as Cultural Inquiry (Nancy
    Duxbury et al.), 10
cultural sector
  definitions, 109–110
  research, 110–113, 118
  in social and economic transformation,
    120–122, 128
  in tourism market transformation, 122–123
  in Yukon, 117–118
cultural tourism
  dynamics, 111–112, 191
  in Indigenous Canadian North, 149–151
  vs. nature-based tourism, 122–123, 124
  in peripheral regions, 129
culture
  in community development, 44, 61, 75
  definitions, 139
  in fragile environments, 140
  revitalization projects, 90, 168
  See also Traditional Knowledge of Elders
"Culture, Sustainability, and Place"
    conference, 11, 286
Cunha, Conceição, 29
Cunningham, James A., 44, 110, 114, 118
Currid, Elizabeth, 262
Curtis, Kynda, 193, 196, 199, 201, 202, 203

D

Dahl, Roald, 213, 214
Danish School of Design, 271
data collection and analysis, 118, 119–127,
    128, 130
Davies, Amanda, 265
Daviet, Sylvie, 110
Dawson City, Yukon, 117, 126–127
Dawson, Jackie

Arctic cruise tourism, 144, 148, 150, 153,
    155, 156
Arctic tourism management, 145, 146, 154
De Beukelaer, Christiaan, 118
de la Barre, Suzanne, 8, 17, 28, 109, 115, 116,
    138, 198
Dear John (film), 243, 245
DEDT (Department of Economic
    Development and Transportation). *See*
    Nunavut, Canada
Deener, Andrew, 265
deep maps and mapping, 18, 211, 218–220
Delind, Laura, 193
Delisle, Marie-Andrée, 34
Den Dekker, Teun, 191, 198, 202
Denis-Jacob, Jonathan, 285
Dennison, Tracy, 172
Destination Bornholm, 270
destination marketing (or management)
    organizations (DMOs), 231, 250–251, 270
destinations
  image enhancement, 229, 249
  promotion, 235–236, 242–244
Detours: A Decolonial Guide to Hawai'i, 9
Di Liberto, Tom, 143
DIG: An Archeological Adventure, 67–68
Dillwyn, Amy, *The Rebecca Rioter*, 217, 219
distinctiveness vs. authenticity, 213–214
DMOs. *See* destination marketing (or
    management) organizations (DMOs)
Dolter, Brett, 112
Dredge, Dianne, 278
Drewes, Ernst, 292
Drolet, Julie, 5
Drummond, Fiona, 79
Drummond, James H., 79, 87, 88
Du Cros, Hilary, 111
Du Pisani, Jacobus A., 9
Dubinksy, Lon, 9
Duif, Lian, 5, 11, 172, 176, 284, 286, 287, 293
Dunlop, Stewart, 62
Duxbury, Nancy
  communities, 10, 74, 81, 140, 265
  conference in Azores, 286
  creative tourism, 4, 8, 16, 23, 28, 210, 230,
    283
  CREATOUR project, 27, 29, 30, 34, 38, 39,
    41, 42, 44, 61, 211

E

ecomuseums, 198
economic development
  and community, 11, 112, 182–183
  correlating features, 120–122
  and cultural development, 9–10, 110, 128
  with festivals, 80–81, 84–85, 91
  with film industry, 242
  with heritage, 67
  impact calculation, 100–102
  in Sweden's policy, 266–267
  with tourism, 2, 61, 62, 74–75
economies
  Canada's North, 141–143
  interdependence with culture, 140
  See also creative economy; cultural sector;
      social economies and systems
Edelson, Natalie, 121
education
  clusters, 68, 69
  models, 72, 73
Einarsdóttir, Gréta S., 171
Eldur í Húanþingi festival, 170
employment opportunities, 171, 180, 183, 242
England, Lauren, 291
entrepreneurs, 39, 44, 115, 183, 203
environment
  climate change, 138, 145, 146
  cruise tourism impacts, 139
  and food tourism, 201
Europe, 5, 62
  See also specific locations
European Photography festival, 74
European Union (EU)
  cohesion policies, 266
  funding and influence, 63–64, 261,
    268–269
EUROTEX project, 290–291
events, 92, 278, 294
  Český Krumlov, 65, 66
  CREATOUR pilots, 40–41
  Härjedalen, 274–275
  Jyväskylä, 69–70
  Óbidos, 71
  Reggio Emilia, 73–74
  Wales Major Events Unit, 213
  Welsh literary, 214
  women's stories in Kluk, 275–276
Everett, Martin G., 264
Everett, Sally, 194, 260, 263
everyday life
  commodification of, 28, 111

escapes from, 82
interactions with, 62, 75
experience tourism, 29, 109, 222
extinction tourism, 145

F

Falconer, Emily, 199
farm shops and markets, 196–197, 202–203
farming
  Iceland, 175
  and tourism, 176–177, 197, 198
festivals, 80–85, 91–92, 100–101
  Český Krumlov, 65
  Dawson City, 117
  European Photography festival, 74
  International Chocolate Festival of Óbidos,
    71
  International Literary Festival FOLIO, 72
  Konstgödning project, 273–274
  Legends of House Festival, 89
  60s Festival, 88–89
  Welsh authors, 213, 214
  See also Mahika Mahikeng Music and
    Cultural Festival
film and tourism collaboration
  advocacy and lobbying, 236–238
  case studies, 231–232, 241
  community development, 244–245
  destination profile-raising, 242–244
  industry and economic development, 242
  marketing, 234–236
  opportunities in, 232–234, 254–255
  organizing for, 248–249
  resource sharing, 238–241
  working relationships, 247
  See also film tourism
film tourism
  creative product development, 245–246
  opportunities, 229, 254–255
  planning, 250–254
  research, 230
filmmaking, 229, 231
fishing industry in Iceland, 175
Fitjar, Rune Dahl, 262
Flannery, Wesley, 171
Fleming, Rachel C., 110
Florida, Richard, 3, 110, 114, 181, 261, 262,
    285, 288, 292
Fløysand, Arnt, 259, 263
food

as creative consumption, 199–201
as creative production, 17, 196–198
regional products, 40–41, 175
tourists' interests in, 193, 198–199
food tourism
as agent of change, 202–203
in CREATOUR pilots, 40–41
International Chocolate Festival of Óbidos, 71
scholarship, 192, 202, 203–204
sustainability of, 193–194, 201
Foster, Don, 181
Fotoatelier Seidel, 65
Foundation Reggio Children, 73
Franklin, Sir John, 143
Frey, Oliver, 183
Friedman, Gabriel, 144
funding
European Union (EU), 63–64, 261, 268–269
film company and tourism office partnering, 237
Mahika Mahikeng Festival, 102
Yukon arts and culture, 123–125, 126, 129

G

Gabe, Todd, 110
Gadwa, Anne, 288
Galloway, Susan, 62
Garner, Alan, *The Owl Service*, 217, 220
Garnham, Nicholas, 62
Garrett-Petts, W. F., 9, 10
George, Jane, 148
George, Wanda, 250
Gibson, Chris, 112, 114, 115, 118, 119, 259, 262, 263, 266
Gil, Sergio Moreno, 230
Gillmor, Desmond A., 165
Gilmore, James H., 209, 222
globalization, 286, 292, 293
Glusac, Elaine, 1
Gonçalves, Alexandra R., 36
Goodwin, Harold, 7
governments
department structure, 248–249
in event organization, 89–91, 211, 212–213
lobbying of, 237–238
re-designation of tourists, 293
rural development promotion, 263

See also Nunavut, Canada; public-private partnerships; Welsh government; Yukon, Canada
Granovetter, Mark S., 264
Granquist, Sandra. M., 171
Grasseni, C., 197
Grettir the Strong project, 168
Griffiths, Niall, *Sheepshagger*, 217, 220
Groulx, Mark, 15
Guy, Emmanuel, 143, 144
Guy, Jack, 243

H

Haalboom, Bethany, 142
Hahm, Jeeyeon, 230
Haila, Yrjö, 122
Hall, Colin M., 115, 166, 193, 199, 201
Hall, Derek, 114, 115, 251, 253
Halvarson, Bodil, 274–275
Handbendi Brúðuleikhús puppet theatre, 170, 180, 182
Happy City (Charles Montgomery), 285
Härjedalen, Sweden, 274–275
Harland, Tony, 119
Harmer, Devin, 89
Harris, John, 213
Harvey, David C., 262, 264, 266, 278
Hauksson, Erlingur, 171
Hautamäki, Antti, 69
Hawai'i, 197, 202
Hawkins, Harriet, 262, 264, 266, 278
Headland, Robert K., 143
Healy, Noel, 171
heat maps, 94, 100
Heitmann, Sine, 241, 244, 250–251
Helgadóttir, Guðrún, 165
Henson, Spencer, 193, 201
heritage, 17, 67–68, 91–92, 210, 283, 290
heritage sites
Český Krumlov, 64–66
York, 66–68
Hewison, Robert, 67
Hicks, Jack, 142
Higginbotham, John, 151
Hildreth, Jeremy, 284
Hills Strategies Inc., 117, 147
HMS Erebus and *HMS Terror*, 143
Hoag, Hannah, 141, 146
Hobden, Karen, 198
Holtkamp, Chris, 176

Hopper, Tristin, 147
hospitality services, 200
Howkins, John, 194, 198, 200, 201, 202
Huang, Yu-Chih, 287
Hudson, Simon, 230, 238, 246, 248, 249, 251
Huggins, Robert, 110
Hughes, Tristan, *Revenant*, 217, 224–225
Huijbens, Edward H., 176
Hull, John, 109, 116
Humphreys, John, 193, 201
Húnaþing vestra, Iceland, 166–169, 175,
    176–177, 179–180, 181–182
Húnavatnshreppur, Iceland, 166
Hvammstangi, Iceland, 167, 171, 173, 174, 175,
    179–180
hyvinvointi (wellness), 69

I

Ice Pilots NWT (television), 246
Iceland, 165, 166, 170–174, 172
Iceland Magazine, 172
Iceland Textile Center, 170
Icelandic Seal Center, Selasetur Íslands, 166,
    168–169, 171
    development of, 177–178
    goals and impacts, 179, 180–181, 183–184
    management of, 182
IdeaLabs, 35, 38, 42
identity
    community support for, 185
    in creative tourism, 191–192, 198
    through food culture, 17, 199–200
    and relationships, 203
    See also sense of place
Ilbery, Brian, 193
illiqusiq (the way it was), 149–150
image enhancement
    with film production collaboration,
        234–235, 242
    vs. reality improvement, 284, 295
immigrants, in art co-production, 274–275
Indigenous communities
    empowerment, 109, 112, 128–129, 130
    knowledge and decision-making, 146, 152
    way of life and tourism, 139, 140, 147,
        149–151
Ingilínardóttir, K. D., 170
Ingold, Tim, 221
innovation, 69–70, 73, 74
International Chocolate Festival of Óbidos, 71

Inuit society, 149–150, 152–153
Ioannides, Dimitri, 112, 118, 204, 247, 259,
    260, 262, 265, 266, 267, 268
Ireland, festivals, 85
Irimias, Anna, 230, 249
Isbell, Matthew G., 249
Ivanovic, Milena, 85, 102

J

Jack, Susan, 119
Jacobs, Jane, 284
Jakobsen, Stig-Erik, 259, 263
Jameson, Frederic, 215
Jämtland, Sweden, 267, 268, 272
Jansson, Johan, 119
Jayne, Mark, 6, 260, 262, 263, 278
Jeannotte, M. Sharon, 28, 137, 140, 152
Jelinčić, Daniella Angelina, 28
Jensen, Jason L., 231
Jóhannesson, Gunnar Thór, 165, 176
Johansson, Martin, 275–276
Johnes, Martin, 212
Johnson, Jeffrey C., 264
Johnson, L. R., 168
Johnston, Adrianne, 145, 148
Johnston, Margaret, 148, 150, 153, 155
Jorvik Viking Centre, 67
Jøsendal, Kari, 262
Jurjonas, Matthew, 185
Juskelyte, Donata, 229, 230, 243
Jyväskylä, Finland, 68–70

K

Karlsdóttir, Guðný Hrund, 173, 180, 183
Kassam, Ashifa, 148
Kastenholz, Elisabeth, 29
Kim, Samuel Seongseop, 241, 244
Kim, Sangkyun, 17, 230
Kindon, Sara, 230
Kloes, Gudrun M. H., 168, 175, 176, 178
Kluk, Sweden, 275–276
Kneafsey, Moya, 263
knowledge
    co-creation and exchange, 35–36, 38, 44,
        168, 179
    local and traditional, 16–17, 40–41, 170
    tourists' consumption, 201, 202
    Traditional Knowledge of Elders, 140, 146,
        152
    urban vs. rural stigma, 171, 266

See also data collection and analysis
knowledge hubs, 69, 73
Kokorsch, Matthias, 174
Kolirin, Lianne, 243
Konijnendijk, Cecil, 172
Konstgödning (Art Fertilization) project, 261,
    267, 272–276, 277
Koschmann, Matt, 249
Kristjánsson, Jón Þ., 171
Kristofersson, Dadi Mar, 175
Kujawinski, Peter, 152
Kung, Shiann-Far, 62, 175, 287
Kyle, Kate, 147

L

Lake District, England, 262
Land, Chris, 91
Landriault, Mathieu, 137
Landry, Charles, 261, 262
Landsbankinn, 171
Larsen, Jonas, 209
Larsen, Karin T., 268
Lash, Scott, 109
Lasserre, Frédéric, 143, 144
"last chance" tourism, 145
Laugarbakki, Iceland, 167, 168, 179
Leaf, Wanda, 117, 129
Leblanc, Pierre, 146
LED (local economic development). *See*
    economic development
Lee, Anne H. J., 17
Lee, Seokho, 85
Lefebvre, Henri, 284
Legends of House Festival, 89
Lemelin, Harvey, 140, 145
Ler Devagar, 72
Leriche, Frédéric, 110
Levy, Stuart E., 244
Lew, Alan A., 81, 82, 85, 169
Lewis, Laurie, 249
Life in New Mexico (television), 239
Lindblad Explorer, 144
*Literary Atlas: Plotting English Language
    Novels in Wales*
    deep maps, 219–220
    novel selections, 215–218
    overview and goals, 211, 215, 226
    resources, 220–221
    tourist reactions, 222–225
literary geography, 214–215

literary tourism, 72, 222–225, 226
Literature Wales, 214
Liu, Shuwen, 195, 200, 203
local economic development (LED). *See*
    economic development
Localhood campaign, 7
locals. *See* residents
Lockie, Stewart, 195
Long, Lucy, 196, 201, 203
Luckman, Susan, 44, 115, 259, 266
Luh, Ding-Bang, 62, 175, 287
Luna, Marcos, 171
Lund, Katrín Anna, 165
Lundberg, Christine, 239, 244, 246
Lune, Howard, 231

M

M&E Framework (Framework for the
    Monitoring and Evaluation of Publically
    Funding Arts, Culture and Heritage),
    80, 92
Ma, Ling, 81, 82, 85
McCoy, Jeremy, 145
McGranahan, David, 262, 265
McKercher, Bob, 111
McKie, Robin, 145, 148
MacLennan, David, 10
McLucas, Clifford, 18
MacPherson, Ian, 142, 143
Maher, Patrick, 109, 116, 146
Mahika Mahikeng Music and Cultural
    Festival
    economic impacts, 100–102
    goals and development, 89–91
    opportunities for growth, 103–105
    overview, 79–80
    valuation study, 92–99
Mahikeng, South Africa, 79, 86–89
Mair, Heather, 184, 250
Make it York, 68
Mandic, Ante, 229, 230, 231, 236, 241, 242, 250
Manniche, Jesper, 268
manufacturing site visits, 71
mapping and maps
    cultural resources, 215, 226, 253–254
    deep maps, 211, 218–220
    festival attendees, 94, 100
    film tourism, 235–236, 243, 253
Margulies-Breitbart, Myrna, 114
marketing

Creative Óbidos campaign, 71
of festivals, 82, 103
film production and tourism
    collaborations, 234–236, 239, 242
in food tourism, 200
nature- vs. culture-oriented, 122–123
and product development, 41–42
Markusen, Ann, 181, 288
Marques, Lénia, 8, 195, 202, 203, 204, 287,
    289, 293
Mashigo, Lehlononolo, 91, 102
Massey, Doreen, 218
Matunga, Hirini, 152
Mayes, Robin, 111, 115
Meethan, Kevin, 67
mega-events, 213, 214
Mercer, Keith, 143
Merriam, Sharan, 119
microfictions, 221
Migdal, Alex, 145
Milne, S., 140, 150
Minguzzi, Antonio, 62
missionary travel, 200
Mitas, Ondrej, 195, 204
Mitchell, Clare J., 260, 263
Mizusawa udon noodle, 17
Mommaas, Hans, 140, 262
Moretti, Franco, 215
Morgan, Jeffrey, 209
Morgan, Kenneth O., 211
Morgan, Nigel, 239, 244, 246
Moscardo, Gianna, 15
Mostafanezhad, Mary, 197, 202, 203
Müller, Dieter K., 109
multiplier effects, 100–101
Murray, Catherine, 44
Murray, Nick, 147, 148
museums, 177, 198, 223
MV Clipper Adventurer, 146

N

Nadotti, Loris, 100
Narcos (television), 243
Natcher, David, 142
nature-based tourism, 122–123, 128
    See also Icelandic Seal Center
Neatby, Leslie H., 143
Nel, Verna, 88
Nelson, Ross, 44
Neto, Maria João, 71

networking
    community-first relationships, 44–45
    and creative development, 263–265, 278
    as creative project, 272–276
    extra-local, 271
    for startups, 38, 39–40, 42
networks
    CREATOUR project, 30, 35, 38, 39, 42
    Óbidos and pan-European, 71
    Sauna from Finland, 69
    Seal Travel, 169
    See also Arts and Crafts Association
        Bornholm; clusters; film
        and tourism collaboration;
        Konstgödning (Art Fertilization)
        project
New Mexico Tourism Department, 239
Nichols Clark, Terry, 285
Nijkamp, Peter, 62
Norcliffe, Glen, 264
Nordic Arctic culture, 110–111
Nordic Council of Ministers, 110
Nordicity Group, 140
North Sea Jazz Festival, 83–84
Northwest Passage, 137, 143–145
Northwest Territories (NWT), Canada, 235
Nunavut, Canada
    cruise ship visits, 144
    marine tourism governance, 154
    planning processes, 152–154
    tourist spending in, 147–148

O

Óbidos Castle, 70–71
Óbidos Literary Town project, 72
Óbidos, Portugal, 70–72
OECD, 31, 109, 210
Officina delle Arti, 74
Ogilvie, Sheilagh, 172
Oksanen, Kaisa, 69
O'Regan, Tom, 230
Östersund, Sweden, 272
Our Common Future (Brundtland Report), 9
The Owl Service (Alan Garner), 217, 220

P

Page, Stephen J., 66
Palmer, Robert, 213
Pantzar, Mika, 194, 288
Parnell, Susan M., 86, 87

Paschinger, E., 286
Pauktuutit (Inuit Women of Canada), 153
Pelly, David, 151
Penny, Meyers Norris, 121
peripheral regions. *See* creative peripheries;
    remote areas
Perri, Giovanni, 199, 201
Petersen, Tage, 268
Petric, Lidija, 229, 230, 231, 236, 241, 242, 250
Petridou, Evangelia, 112, 118, 119, 204, 247,
    259, 260, 262, 265, 266, 267
Petrov, Andrey, 110, 112, 114, 115, 117, 118,
    130, 147
Phillips, Rhonda, 181, 185
photography, 40, 65, 74
PIC framework, 252–253
Pimpão, Marta Beatriz Oliveira, 71
Pine, Joseph P., 209, 222
Piracha, Awais, 140
Pivcevic, Smiljana, 229, 230, 231, 236, 241,
    242, 250
placemaking
    elements, 287–288, 291–292
    historical treatment, 284–285
    at Mahika Mahikeng festival, 90, 105
    opportunities, 11, 112, 120, 176
    vs. place marketing, 284
places
    attractiveness of, 285, 295
    distinctiveness, 115
    festival locations, 82–85
    film locations, 235–236, 254
    and products, 175–176
    real-and-imagined, 215, 225
    See also destinations; remote areas; rural
        areas; sense of place
planning
    community engagement, 9–10, 184–185,
        230, 250–251, 252, 254
    and Indigenous world views, 152–154
    PIC framework, 252–253
Pliner, Patricia, 198
plotpoints and plotlines, 219–220
political power and engagement, 171
Pond Inlet, Nunavut, 144, 147, 151
Portugal, 27–28, 39, 70–72
postmodern consumerism, 209
Power, Dominic, 119
Pratt, Andy, 109, 261, 262
Pratt, Stephen, 241
Price, Karen, 214

Prince, Solène, 112, 118, 204, 247, 259, 265
private and public partnerships. *See* public
    and private partnerships
products
    development, 41–42, 66, 71, 148, 151, 246
    and place, 175–176
    prohibited, 147
property improvement, 239, 245
public input
    CREATOUR project pilot selection, 37–38
    Mahika Mahikeng Festival, 90–91
public-private partnerships, 68, 69–70, 71–72,
    168–169, 171, 278
Puren, Karen, 292

Q

Qaujimajatuqangit, Inuit belief system,
    152–153
Quinn, Bernadette, 85, 102

R

Rana, Ratna, 140
Rantala, Outi, 109
Ray, Christopher, 259, 261, 263
Raymond, Crispin, 4, 5, 28, 288, 290
Rebecca Rioter, The (Amy Dillwyn), 217, 219
Recher, Vedran, 229, 230, 241, 244, 250
Reggio Emilia, Italy, 72–74
Reid, Donald G., 184, 250
remote areas
    and creativity, 114–116
    See also Arctic regions
Rendace, Olivero, 264
A Research Agenda for Creative Tourism
    (Nancy Duxbury and Greg Richards), 8
research tools
    case study approach, 119
    "talking whilst walking," 221
residents
    as art project participants, 273–276
    impacts of tourism on, 7–8, 66, 224–225,
        244–245
    See also communities; visitor-resident
        interactions
resilience, 166, 169, 176, 184
Revenant (Tristan Hughes), 217, 224–225
Reykjavík, Iceland, 171, 172
Reynarsson, Bjarni, 166
Rhys-Taylor, Alex, 221
Richards, Greg

advisor, 34
communities, 5, 74, 81, 172, 176, 181, 210,
   215, 251, 254
creative tourism, 4, 8, 16, 43, 140, 192, 226,
   230, 287, 288, 289, 291
creative turn, 28, 111, 209
creativity, 198, 202, 261
cultural tourism, 89, 90, 122, 191
Finnish craft skills, 290
industry and economy, 110, 116, 183, 194,
   203, 259
mega-events, 213
places and placemaking, 11, 229, 236, 260,
   262, 283, 284
tourists, 6, 23, 62, 165, 175, 209, 286, 293
Ritchie, J. R. Brent, 244
Ritsema, Roger, 153
Roberts, Elisabeth, 176
Roberts, Les, 243, 244, 253
Robertson, Kent, 5, 6
Rodgers, Robert, 231
Rodon, Thierry, 151
Rogerson, Jayne M., 89
Rojek, Chris, 213
Romão, João, 62
Rudy, Jill T., 200, 203
rural areas
   as art co-production space, 272–276, 277
   creative industry development in, 259–260,
      262–263
   networking in, 264–265, 276–278
Russo, Antonio Paolo, 6, 289

S

Saarinen, Jarkko, 166
Saayman, Andrea, 85
Saayman, Melville, 85
St Roch, 143
Sæthórsdóttir, Anna Dóra, 166
Sauna from Finland, 69
Saunders, Ahmed, 221
Scherf, Kathleen, 1, 74, 80, 192, 226, 230, 260,
   276, 289
Schnell, Steven, 193, 194
Schrock, Greg, 181
Schuster, Thomas J., 85
Scott, Allen J., 110, 261, 262, 264, 278, 288
sea ice losses, 143, 146
seal products, 147–148
seal tourism, 168, 169, 178–179

Seal Travel, 169, 179
Seaman, Bruce, 85, 101
Seekamp, Erin, 185
Seidel, Josef and František, home and studio,
   65
Selasigling, 178
Selby, Martin, 67
Selin, Helaine, 122
sense of place
   concepts, 195
   in food tourism, 196, 199–200, 201,
      203–204
Sevunts, Levon, 156
Seyðisfjörður arts village, Iceland, 170
Seyfang, Gill, 193
Seyfried, Amanda, 244
Sharp, William, 214
Sheehan, Rebecca, 265
Sheepshagger (Niall Griffiths), 217, 220
Shenoy, Sajna, 198
Shifts (Christopher Meredith), 217, 222–223
shipping and ship traffic, 144–145, 146
   See also cruise tourism
Shove, Elizabeth, 194, 288
Sigurjonsdottir, Hrefna, 171
Silva, Sílvia, 30, 34, 41, 42
Sims, Rebecca, 263
Sirhowy Ironworks, 222
sites
   heritage, 64–66, 66–68
   threatened and sensitive, 146, 150, 154–155
60s Festival, 88–89
skills sharing and development, 19–17, 65,
   289–290
Slocombe, Scott, 146
Slocum, Susan L., 191, 193, 194, 196, 198, 199,
   201, 202, 203
Small Cities with Big Dreams (Greg Richards
   and Lian Duif), 286
smaller communities
   features, 5–6, 293
   as film locations, 251–252
   vs. larger communities, 170, 172–173,
      285–286
   tourism potential, 1–2, 10, 23, 27, 63, 210
   tourism strategies, 291–292, 294
Smed Olsen, Lise, 112
Smith, Karen, 230
Smith, Kieron, 209
Smith, Melanie, 139, 149, 150, 151
Smith, Stephen L. J., 119

Snowball, Jeanette, 79, 85, 89, 90, 101
Soares, Clara Moura, 71
social attitudes, 209
social development
    in creative and cultural sector, 85, 112,
        120–122, 128
    in creative peripheries, 114–115, 266
    integration of immigrants, 274–275
    Jyväskylä experience, 69–70
    in M&E Framework, 92–93
    Mahika Mahikeng festival experience, 90,
        98–99
social economies and systems
    Canada's North, 141–143, 146, 156
    interdependence with culture, 140
Soja, Edward, 215
Solima, Ludovico, 62
Solnit, Rebecca, 222
Sorensen, Chris, 147
Sorensen, Tony, 262
South Africa, tourism, 100
South Carolina films and tourism, 243–244,
    245
Southcott, Chris, 109, 141, 142
space
    concepts of, 195, 284–285
    creative, 260, 270–271
    social, 84
sports events, 65, 142, 272
Spotted by Locals, 289
Stake, Robert, 231
stakeholder theory, 250
start-up support, 42, 44
State & Main (film), 255
Steinholtz, Vanja, 273–274
Stewart, Emma, 145, 146, 148, 149, 150, 153,
    155
Stoecker, Randy, 119
Stone, Kirk, 173
Stratton, Jon, 263
Struzik, Ed, 145
Strydom, Wessel, 292
Sung, Heekyung, 181, 185
sustainability
    of communities, 211
    vs. resilience, 169
sustainable development, 9
sustainable tourism
    and community, 183
    concept, 138–139
    development, 42, 66, 193–194

practices, 74–75, 201
    public and private partnerships, 71–72
Sweden, 266–267, 274

T
Tagalik, Shirley, 153
Taggart, Malcolm, 117, 129
Tan, Siow-Kian, 62, 175, 287
Tanzi, Christine, 243
Tatum, Channing, 244
technology sector, 68, 69–70
territorial cohesion, 266, 267, 272, 277
thematic years, 213
Thomas, Dylan, 213
Thomas, Nicola J., 262, 264, 266, 278
Thomlinson, Eugene, 229
Tibere, Laurence, 199, 203
Timothy, Dallen J., 252, 254
Tkalec, Marina, 229, 230, 241, 244, 250
Toerin, Daan, 100
Tomaz, Elisabete Caldeira Neto, 61, 67, 81,
    90, 231
Þorbjörnsson, Jóhann. G., 171
Törnqvist, Gunnar, 119
Tosun, Cevat, 252, 254
tourism
    European industry, 62
    experience tourism, 29, 109, 222
    extinction (last chance) tourism, 15, 145
    nature and dependencies, 194–195
    research approaches, 119, 221
    trends, 1
    See also creative tourism; cruise tourism;
        cultural tourism; film tourism; food
        tourism; sustainable tourism
tourism development
    community engagement in, 9–10, 75,
        184–185, 230, 250–251, 252, 254
    economic goals, 2, 61, 62, 74–75
    urban bias in, 39, 114
Tourism Industry Association of Canada, 143
tourism workers, 148, 169, 239, 242
tourists
    as change agents, 140, 203
    as creators, 221
    and food, 198–201
    insensitivities of, 150
    interests and values of, 7, 194, 209, 263,
        286, 287
    on Literary Atlas field trips, 222–225

relationship with place, 22–23, 293–294
spending by, 147–148
transformative experiences of, 202
Townsend, Leanne, 176
Traditional Knowledge of Elders, 140, 146, 152
traditions, 149, 194, 198, 290–291
    See also heritage
Tranter, Emma, 144
Tredegar, Wales, 222–223
Tregear, Angela, 193, 264
Tung, Vincent Wing Sun, 230
Turismo Portugal, 27

U

Ulukhaktok, Northwest Territories, 147, 149
UNESCO
    on culture and tourism, 139, 140
    designations and lists, 64, 66, 68, 72
United Nations, 2, 9
UNWTO (United Nations World Tourism
    Organisation), 1, 9, 139, 199, 290
Uqsiq Communications, 140
urban centres, 6–8, 195, 262, 285, 288
urban-centric biases
    in policies and perceptions, 170–171,
        263–264, 265–266
    in research and assessments, 39, 114, 259
urban development
    and creative industries, 62–63
    UNESCO Creative Cities designation, 68,
        72
urban-rural migration, 2, 170, 180
Urry, John, 67, 109, 191, 209

V

Valli del Bitto, Italy, 197–198
Van Aalst, Irina, 81, 83, 84
Van Melik, Rianne, 81, 83, 84
Van Winkle, Christine, 229
Van Zyl, Ciná, 80, 81, 82
Vannoni, Valeria, 100
Verbeek, Desirée, 140
Vercoe, Richard, 180
Verdich, Madeleine, 115
Vergunst, Jo Lee, 221
Viglundsdóttir, Hrafnhildur Yr, 176, 177–178,
    183
Viking life experiences, 67
Vinagre de Castro, Tiago. See Castro, Tiago
    Vinagre de

VÍS (insurance company), 171
Visit Albuquerque, 235
Visit Britain, 243
Visit Wales, 211, 212, 213
visitor guides, 122–123, 124
    See also mapping and maps
visitor-resident interactions
    benefits of, 62–63
    as equals, 22, 203, 289–290, 291, 293–294
    for sustainable tourism, 151
Visser, Gustav, 80, 81, 84
Volo, Serena, 230, 249
volunteering, 197

W

Waitt, Gordon, 114, 262
Wales, 6, 211–213, 226
    See also Literary Atlas: Plotting English
        Language Novels in Wales
Walker, Kaye, 15
Walker, Valoree, 141, 142
walking, 221, 222
Wang, Youcheng, 230
Ward, S., 140, 150
Ward, Susan, 230
Watson, Allan, 262
Watson, Matt, 288
Weatherell, Charlotte, 193
Weaver, Desirée, 138
Weidenfeld, Adi, 62
wellness tourism, 69
Welsh Government (WG), 211, 212–213
Wenzel, George, 140, 150, 152
Westfall, Rachel, 121
WG. See Welsh Government
White, Pauline, 264
Whitehorse, Yukon, 116–117, 126–127
wildlife
    and sea ice loss, 146
    See also seals
Williams, Lindsay, 230
Williams, Raymond, 139
Willoughby-Smith, Julie, 115, 259, 266
Willson, Margaret, 165, 174
Wilson, Julie
    on creative industries, 203, 229, 236, 254,
        261, 262
    on tourism, 62, 191, 209, 226
Wisansing J., 286
Wojan, Timothy, 262, 263, 265

women's stories project, 275–276
World Tourism Organization, 111
Wray, Meredith, 229, 244, 249, 251, 254
WWOOF (World Wide Opportunities on
    Organic Farms), 197

X

Xiao, Honggen, 119

Y

Yin, Robert, 119
York Archaeological Trust, 67
York, England, 66–68
young people, 170, 290
Yukon, Canada
    arts sector, 123, 125–127
    brief overview, 113
    case study data sources and analysis, 119,
        128–130
    communities, 116–117
    cultural sector performance, 117–118
    transformational correlations, 120–122
    visitor guide analysis, 122–123, 124

Z

Zanasi, Luigi, 117, 129
Zasada, Ingo, 263
Zerehi, Sima, 148
Ziakas, Vassilios, 239, 244, 246
Zilic, Ivan, 229, 230, 241, 244, 250
Zukin, Sharon, 181, 262
Žuvela, Ana, 28